HANDBOOKS OF AMERICAN NATURAL HISTORY

Mammals of the Eastern United States

SECOND EDITION

Handbook of Salamanders
Sherman C. Bishop

Handbook of Turtles
Archie Carr

Handbook of Nature Study
Anna Botsford Comstock

The Spider Book
John Henry Comstock

Handbook of Waterfowl Behavior
Paul A. Johnsgard

Aquatic Plants of the United States
W. C. Muenscher

Handbook of Lizards
Hobart M. Smith

Handbook of Frogs and Toads of the United States and Canada
Albert Hazen Wright and Anna Allen Wright

Handbook of Snakes of the United States and Canada
Albert Hazen Wright and Anna Allen Wright

Bull moose, *Alces alces*. Drawing by W. C. Dilger.

MAMMALS

of the Eastern United States

SECOND EDITION

William J. Hamilton, Jr.
Cornell University

and

John O. Whitaker, Jr.
Indiana State University

COMSTOCK PUBLISHING ASSOCIATES a division of

CORNELL UNIVERSITY PRESS / Ithaca and London

To our wives,
Nellie and Royce

Contents

Preface

The land mammals naturally inhabiting the states east of the Mississippi River are briefly described in this volume, along with accounts of their habits and distribution. Like the first edition, this book is designed for students of mammalogy, for professional conservationists, and for the increasing number of laymen who are interested in wildlife. Although there are many good state lists that discuss the distribution and habits of mammals, there continues to be a need for a more comprehensive and detailed work on eastern mammals. This book, which includes illustrations of the species, distributional maps, and important references, will, we hope, help to fill this need. Perhaps too it will suggest to the amateur naturalist or the laboratory professional the joys of collecting and the value of further field study of our mammals, of which little is yet known. Also, it should be useful to those interested in ensuring the survival of species currently considered endangered or threatened.

We have been fortunate to collect mammals in all of the twenty-seven states east of the Mississippi River, except Mississippi. We have visited the high reaches of Mount Katahdin in northern Maine, where water shrews and moose hold company, and the cypress swamps of lower Florida, where the spoor of the mountain lion may be seen. We have trapped spermophiles and prairie voles in Illinois and Indiana; in the Kentucky mountains, wood rats and pine mice have blundered into our traps; and rare bats, after countless misses, have fallen to our guns. In addition, we have spent considerable time studying bats in the caves and mines of Illinois, and rodents and insectivores in the mountains of North Carolina.

We are indebted to a host of workers, most of whom must be left unmentioned. Many pleasant days have been spent studying the huge collections in the United States National Museum, the American Museum of Natural History, and the Philadelphia Academy of Sciences. We examined collections in the Chicago Field Museum, the Chicago Academy of Science, and the Museum of Comparative Zoology at Harvard. We are pleased to be able to provide pictures that illustrate the distinguishing characteristics of almost all the species. Most of the illustrations are appealing as well as informative; a few that do not reproduce well were included because they are rare illustrations or because they show especially important characteristics. Both Arthur Smith and Douglass Payne of Cornell University helped in photographing captive animals. Earle L. Poole contributed his incomparable illustrations, which add measurably to any merit the book may have. In addition, we appreciate very much the discussions, ideas, and help given by our graduate students over a period of many years. Their advice has greatly affected the material in these pages. We must also acknowledge the excellent work of our typist, Laura Bakken, who, by her constant attention to detail and consistency, has greatly improved the text. For their helpful editorial work after the manuscript had been finished, we must thank Lisa Turner and Daniel Snodderly of Cornell University Press.

W. J. HAMILTON, JR.

J. O. WHITAKER, JR.

Ithaca, New York, and
Terre Haute, Indiana

Mammals of the Eastern United States

SECOND EDITION

Introduction

To define a mammal, one need only state that it is an animal more or less covered with hair at some time in its life. Even the great whales are not entirely devoid of hair. In addition, the young of mammals are fed milk produced by mammary glands of the female. Mammals occur everywhere in the eastern United States, and they range from the peaks of the highest mountains to plowed fields that may appear completely barren of vegetation.

Why study mammals? Many of our eastern species, such as the meadow vole, the pine mouse, the cotton rat, and at times the cottontail rabbit, are important pests of agriculture. In order to protect our plants and to prevent the spread of disease, we must learn all that it is possible to know about the habits of such animals. If we can discover the weak link in the life history of the pest, we can reduce it through this channel.

Aside from economic considerations, the study of our native mammals is of absorbing interest and holds a fascination quite different from that of bird study. To see a red fox hunting mice in the bare March meadows, or to uncover the nest of lustrous-eyed flying squirrels—yes, even to come upon the serpentine march of a skunk family in the still June evening—provides a thrill occasioned by no other group of animals. For we ourselves are mammals, and our closest kin have much in common with us.

Taxonomic Treatment: Species and Subspecies

We have included 105 species of terrestrial mammals in this volume. Our thinking on species differs greatly from that of most authors, partly because the emphasis in this work has been on the habits of mammals, but primarily because we feel an excessive number of subspecies has been described for eastern forms. Subspecies, and originally species also, were traditionally described on the basis of morphological differences alone. However, the biological concept of reproductive isolation has replaced the morphological concept at the species level. Whitaker (1970) proposed that a biological concept now be instituted for the recognition of subspecies as well. He believes that subspecies should serve, as the word suggests, for the situation where populations or groups of populations are on the way to becoming new species. The first step in speciation is the institution of a primary isolating mechanism (the stopping of gene flow between the groups of populations involved). Primary isolating mechanisms are generally geographic in nature, that is, groups of populations are separated spatially. When primary isolating mechanisms are in effect, the development of secondary isolating mechanisms can begin. Secondary isolating mechanisms are factors that will prohibit interbreeding once the primary isolating mechanisms have broken down, that is, once members from the two groups again occur together. If secondary isolating mechanisms have developed to the point where the two forms remain distinct, even when occurring together, we say that speciation has occurred. The biological subspecies concept proposed by Whitaker is that the first step in speciation, the development of a primary isolating mechanism, be used as the subspecies criterion. In other words, if we can show that gene flow is being prevented between two populations, and, as a result, the two groups are undergoing divergent evolution, then we should recognize subspecies.

Morphological variation between series of interbreeding populations is expected, as each separate population evolves to some degree to fit its own unique environment. Such variation will occur gradually or abruptly over the animal's range, depending on the extent of environmental variation. Many of the presently described subspecies appear to be based on this sort of variation. Such variation should be recorded, but, in our opinion, should not be used as the basis for describing subspecies.

We believe that subspecies should be retained for only those groups of populations which are separated by primary isolating mechanisms. This application would lead to a reduction in the number of subspecies, and thus would greatly simplify the subspecies arrangement of eastern mammals. In this work, however, the emphasis is on habits rather than taxonomy, and we have omitted discussion of subspecies.

Distribution

Anyone who has tried to prepare range maps recognizes the difficulty of drawing proper boundary lines, as too few data are available, and collectors are forever adding to the present known range. For the maps in this book, we have relied on our own judgment, which is probably often faulty, but based on critical examination of thousands of museum specimens and correspondence with other biologists, with the members of game commissions, with museum curators, and above all, with interested amateurs. Occasionally a state line forms the limit of a known range, but this coincidence is usually explained by the presence of an impassable river barrier. Wild animals are no respecters of political boundaries; they follow natural ecological associations of which we are often unaware and which we cannot fully understand. The maps may be misleading within a few decades, as many species are extending their ranges, for example, the western harvest mouse, the opossum, and the gray fox. On the other hand, ten years could see the virtual extirpation of the gray wolf in its last eastern stronghold, or the extermination of the Florida mountain lion and Leib's bat.

The reader cannot help but note that many northern species follow the Allegheny Mountains through Virginia and West Virginia, finally occurring at their southernmost range in the Great Smoky Mountains of North Carolina and Tennessee. There are species that likewise shun the higher levels, preferring the coastal marshes or sandy beaches of the Atlantic Coast. Quite possibly, there are insular races yet to be described.

Coloration

Mammals often vary much in color; it is thus hazardous to base the description of a new subspecies on coat color alone, as has often been done in the past, particularly if only a few specimens are available. A striking instance of the variability of color within a single subspecies is shown in the accompanying illustration (Figure 1) of the southern fox squirrel. Here we find three distinct color phases, yet the specimens drawn here were collected within a few miles of one another.

Collecting Mammals

Small mammals are for the most part of a shy and retiring disposition; often they are nocturnal and dwell beneath the matted grass or the thick humus cover of the forest. To be sure, there are some, such as the fat-bellied woodchuck of the clover fields and wild cherry hedgerows, or the various ground and tree squirrels, that are readily observed. But with many species we must capture specimens and observe their habits under captive conditions if we are to study the details of their lives that cannot be learned from field observation. Only in this manner can the molt be determined, or gestation, postnatal development, and other reproductive habits be recorded, and much else of

Figure 1. Variation in fox squirrels, *Sciurus niger.* Drawing by Robert J. Lambert.

interest learned which would be quite impossible if captive individuals were unavailable.

Capturing animals alive and uninjured and marking them in sundry manners, either by ear or ankle tagging, ear punching, or toe clipping, do permit recognition of the individual when it is released and recaptured. These methods have been used with increasing success during recent years. Rabbits, squirrels, mice, shrews, bats, and even the larger fur bearers have thus been studied, and a great deal has been learned of their habits.

Animals must be taken in traps designed to kill them if large numbers are needed for systematic study, or if a biological reconnaissance of the mammal fauna must be made in a limited time. It is a lamentable fact that great series of skins of many mammals are often prepared by field biologists without a thought to recording the obvious facts of the living animals' natural history. Furthermore, we often pass by the commonplace, insisting that there is no need to retain a series when the animal is yet abundant. There is not an extant skin or skull of the Florida wolf, exterminated during the twentieth century!

Small rodents and shrews are readily trapped with the familiar snap-back mouse trap, a half dozen of which can be purchased for a dollar. These small traps are likely to crush the skulls of captured specimens, so wherever possible the larger museum trap, especially designed for museum collectors, should be used; it can be purchased from the Animal Trap Company of Lititz, Pennsylvania. The large snap-back rat trap is useful in capturing squirrels, wood rats, and even weasels; and steel traps of various kinds, particularly those which instantly kill the animal, are most useful. The Conibear trap is particularly effective. The Tomahawk Live Trap Company of Tomahawk, Wisconsin, carries a large assortment of cage traps.

Various baits have proved useful, but these often are as attractive to the ants, crickets, and slugs of a region as to the mammal for which the collector intended them. Consequently the trapper may find many unsnapped traps from which all vestiges of bait have been removed. Peanut butter, bacon, and oatmeal, thoroughly mixed, form a standard bait used by many collectors, although others use peanut butter alone. Suet is attractive to many mammals, particularly during the winter months. Raisins, canary seed, liverwurst, bits of meat, kippered herring, sunflower seed—indeed, the table scraps from one's kitchen—will appeal to most mammals. Or, failing bait, if one places the trap directly across the runway of the animal so that the treadle will be tripped as the animal moves by, many will be caught, for most small mammals seem totally without suspicion. Traps can also be sunk crosswise of subterranean burrows, so that the treadle is at ground level of the burrow. For added effectiveness, the hole can be covered with bark or other material, but it must be high enough that the treadle will clear it.

Post holes or sunken cans are often effective, and will take a variety of small mammals in short order. Where small mammals are numerous, they can be overtaken and caught by hand, but a stout glove should be worn to guard against the sharp teeth. We have collected many field mice in this fashion, and also chipmunks, ground squirrels, and jumping mice.

Shooting is effective, and it is sometimes the only manner in which certain bats can be taken. If the collector is stationed over a watercourse to which bats repair in the evening, the individuals that are shot will fall in the water and can easily be recovered. But to find a bat that has been brought down in the rank grass and weeds in the fading light of a summer evening will tax the patience of a Job, and most specimens are lost. In caves, bats can often be secured merely by picking them from the walls or ceiling, although some species, notably the gray bat, the big-eared bat, and the Indiana bat, are presently being rapidly reduced in number and should not be disturbed in their winter quarters. A pair of long forceps will often prove useful to extract a hidden bat from some otherwise inaccessible fissure.

Japanese mistnets have come into wide usage by bird banders and are also very effective in capturing bats. A greater variety of species may be taken with nets than in any other manner. The use of these nets is illegal in many states, and permits from the conservation department or fish and game commission must be obtained. Nets must be used with care in caves or other areas where many bats will be caught in a short time, as it often takes some time to remove a bat from the net. A good situation for netting operations is over woodland pools where a canopy of branches restricts the area of flight. It is well to observe the flight pattern of bats before arranging the net. After capture, bats may be placed temporarily in a hardware cloth cylinder.

Bat banding has measurably increased our knowledge of chiropteran biology. The small bands placed on the forearm have been carried by some bats for more than twenty years. Bat detectors, which pick up the supersonic notes of bats, can be used to find bat roosts and bat feeding areas. With them we learn about the bats' times of activity and other facets of bat behavior. For those engaged in bat research, *Bat Research News* will prove invaluable. Dr. M. B. Fenton of the department of Biology, Carleton University, Ottawa, Ontario, is the editor.

Fur dealers and trappers, visited in the late fall and winter, can provide valuable specimens. From them, the collector can procure fine skull series, useful measurements of the larger mammals which are otherwise difficult to secure, and valuable information on primeness, reproduction, the habits of hibernation of the two sexes, food habits, internal and external parasites, and other subjects.

The preparation of mammal skins, skulls, and skeletons for museum purposes is well outlined in *Wildlife Management Techniques,* published by the Wildlife Society (1971). This volume contains many techniques for field study and is a mine of information. There is no need to repeat the procedures here. If young collectors observe an experienced collector preparing a skin, they will save much time and disappointment.

One skin with full notes and a well-recorded label is worth a hundred specimens without data. We cannot impress too strongly on the reader the desirability, indeed the utmost urgency, of recording in minute detail everything connected with the collecting of the specimen. It is amazing how much can be written on a museum label measuring but one-half by two inches. By

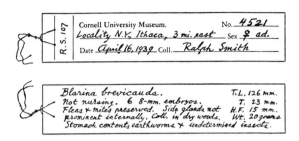

R.S. 107

Cornell University Museum. No. *4521*
Locality *N.Y., Ithaca, 3 mi. east* Sex *♀ ad.*
Date *April 16, 1939* Coll. *Ralph Smith*

Blarina brevicauda. T.L. *126 mm.*
Not nursing. 6 8-mm. embryos. T. *23 mm.*
Fleas & mites preserved. Side glands not H.F. *15 mm.*
prominent internally. Coll. in dry woods. Wt. *20 grams*
Stomach contents, earthworms & undetermined insects.

using a fine pen with India ink, the following data may and should be recorded on such a small tag: locality; date; collector; sex of specimen, with probable age, as "immature" or "adult"; the conventional measurements, as total length, tail, hind foot, and whenever possible, weight; condition of the reproductive organs (if a female is pregnant, the number of embryos, their size and the fact that the female is or is not nursing; if a male, a statement regarding the relative size of the testes); stomach or pouch contents; presence of parasites; details of habitat; and anything else of significance. The name of the specimen is of no immediate consequence: if it is known, record it on the label, if not, it can readily be determined at a later date.

In addition, further notes should be kept in a field journal, in which the specimens are numbered consecutively as they are taken from day to day through the years. These notes should include details on the habits as observed in the field, the conditions under which the specimen was secured, sketches of any soft parts the character of which may be lost in drying; i.e., the position of the teats, the plantar tubercles, the molting pattern, etc. Indeed, no collector should feel concerned lest he keep too many notes; this is a faultless occupation.

How and What to Study

In order to study the details of life history, we often must collect the live animal. The way in which this is done depends on the size and wariness of the species. Many small forms can be caught in the hand, and there are several traps on the market today for taking mammals varying in size from the smallest of shrews to bobcats.

Small mammals may be kept in aquaria. One may build larger cages or purchase one of the many commercial cages presently on the market. There are many sizes and shapes of plastic containers available now which could serve as animal cages. A suitable screen cover will keep the occupant in. Most captive mammals are kept under conditions that are too dry for them. Damp sphagnum, leaf mold, or sod should be placed in a portion of the container, and a suitable nest chamber provided where the captive can retire to sleep, or, if fortune favors, to bear her young.

In the field, direct observation can often be made on feeding animals under normal conditions; there burrows can be excavated and the nests described, the active period of diurnal species recorded, or the feces of predators recovered and examined in the laboratory. These are much-neglected studies in which mammalogists, without recourse to museums and natural history laboratories, can make their own contributions to the science of mammalogy.

What to study? Every species described in this volume would amply repay close study. Of the 105 species treated, the young of many have yet to be described; the gestation period of a number of species is unknown; detailed data on the food habits of relatively few have as yet been recorded. Home life, seasonal differences in food habits, activity cycle, and behavior are unknown in any detail for many of the species. We need to know more about the winter behavior of mammals. Additional information is desirable on hibernation, the migration of bats, and the local wanderings of game species. A proper evaluation of the soft parts and other structural modifications in relation to the economy of the species should be made.

The fleas, lice, ticks, and larger mites infesting many of the species are fairly well known, but only a few hosts have been adequately studied for small ectoparasites. Indeed, new species of mites are still being described from eastern mammals. Use a dissecting microscope to examine the hair and skin. The hair can be manipulated with dissecting needles. External parasites can be preserved directly in 70 percent ethanol. Likewise, internal parasites have been little sought in many of the less common species. Nematodes can be killed in hot 70 percent ethyl alcohol, then preserved in a mixture of four parts 70 percent ethanol and one part glycerine. Other worm parasites can be killed and stored in AFA, a mixture of twenty parts 95 percent alcohol, six parts commercial formalin, one part glacial acetic acid, and forty parts distilled water.

Techniques for determining movements, home ranges, growth, aging behavior, and other phases of the life history of mammals have become greatly refined in the past thirty-five years. When the first edition of this book was written, ear tagging and toe clipping were standard procedures for identifying recaptured animals. Now radiotelemetry, radioisotopes, and other recent developments have aided greatly in our understanding of the habits and behavior of wild animals. Tranquilizers (nicotine salicylate, succinylcholine, and others) shot into the animal with a dart gun have been effective with the larger mammals. Once the animal has been immobilized, it can be tagged in sundry ways.

W. G. Sheldon (1949) put expanding collars on young foxes and secured much information on their movements. Where collars are not practical, a braided polythene rope harness with color patches has proven effective. The large-eared ungulates have been marked with a washer tag holding a colored plastic disc, permitting the animal to be observed at a distance. Commercial fur dyes are also useful in marking animals. Their use is described by F. C. Evans and R. Holdenreid (1943). Such marks persist only until moulting occurs. Freeze-marking mammals with a pressurized refrigerant destroys the pigment cells in the hair follicles and in a few weeks a new growth of unpigmented hairs become evident at the brand site (Lazarus and Rowe 1975).

Gillian K. Godfrey (1954) has used a leg bracelet modified to hold radioactive cobalt so that movements of marked voles could be traced with a Geiger-Muller counter. The rapid advances in electronics have provided new approaches to the field study of wild mammals. Radiolocation telemetry is discussed by Adams (1965). The cost of this tracking equipment is high, but its merits have been demonstrated.

The approximate age of many mammals may be determined by tooth wear. This method has been carefully worked out by Severinghaus (1949) for white-tailed deer. Similar procedures have been established for many carnivores, ungulates, and rodents. Tooth wear in *Sorex fumeus* and some other shrews is distinctive and reliably establishes age classes (Hamilton 1940). The annulations on the root of the upper canine in the black bear appear to be correlated with age. H. T. Gier (1957) has provided a reconstruction of tooth wear in coyotes from one through eight years. This method may undoubtedly be applied to other carnivores. Epiphyseal cartilage is prominent in many of the long bones of immature animals, but disappears as maturity is reached. The baculum has been widely used in an attempt to establish age classes. This method should be used with caution for there is wide variation in the rate of development of this structure.

The eye lens grows continually throughout life, and thus may act as an indicator of age. It has been used with some success. For example, R. D. Lord (1959) used the dried weight of the lens in cottontail rabbits to determine age. The lens-growth curve permits the determination of month of birth of the younger rabbits and the year of birth of rabbits older than one year.

Much interest has been generated in the numbers of a species in a given area. Population trends and the characteristics of population cycles have been studied by many investigators. Their techniques are varied and often complicated. The purpose of population studies is to get as close an approximation of the population as the resources permit. Quite as important, one must state the presumed accuracy of such an estimate. The various methods employed in such studies are well summarized by W. Scott Overton and David E. Davis (1969).

The food and feeding habits of some mammals have been much studied, largely because of economic considerations. The dietary habits of larger species and those smaller forms which have significant interest to the agriculturist have been fairly well explored, but the food habits of most mammals are imperfectly known. For the most part, mammals feed, within limits, upon whatever is available. Thus their food may vary from day to day, and often markedly with the seasons. The usual procedure in food habits research is to examine stomach contents and scats. The investigator must have a thorough knowledge of the species he is studying, for the character of an animal's dropping will vary with the type of food and with the season. It is essential that a good reference collection of vertebrates, invertebrates, and plant material be available for comparison, but even more valuable is a collection of potential food items made at the time and place of collection of the mammals. Few investigators are sufficiently knowledgeable to identify all the plant and animal matter encountered in such a study.

Research material for mammal studies may be obtained in many ways. Road kills are particularly valuable subjects for study. A gallon or two of 10 percent formalin and a supply of single-edge razor blades kept in the car will allow rapid removal and preservation of the desired parts. Carcasses provided by fur trappers are available during the fall and winter months. Trappers are sometimes employed by state game commissions, and arrangements may often be made with these agencies for a cooperative research project.

We have suggested only a few lines of research that show promise. The references listed in the back of the book provide information on the techniques that have proven effective. There are many opportunities for research into the lives of our native mammals that remain virtually unexplored. There is little need to travel to the far corners of the world; unique opportunities exist in our own backyards. Our most common species offer unlimited scope for scholarly studies. Indeed, when one undertakes a survey of the commonplace, their often great numbers make the chances bright for uncovering new concepts.

It is imperative that investigators keep full and detailed notes on their observations. Memory is ephemeral; what we see today is lost tomorrow if not committed to paper. More than a century has passed since the renowned American naturalist Elliot Coues (1872) stressed the utmost necessity for keeping meticulous notes. After a full day in the field, and making observations on a score of different things, Coues stated,

> Now you know these things, but very likely no one else does and you know them at the time, but you will not recollect a tithe of them in a few weeks or months, to say nothing of years. Don't trust your memory, it will trip you up, what is clear now will grow obscure; what is found will be lost. Write down everything while it is fresh in your mind; write it out in full; time so spent now will be time saved in the end, when you offer your researches to the discriminating public. Don't be satisfied with a dry-as-dust item: clothe a skeleton fact and breathe life into it with thoughts that glow; let the paper smell of the woods. There's a pulse in a new fact; catch the rhythm before it dies. Keep off the quicksand of mere memorandum—that means something "to be remembered," which is just what you cannot do. Shun abbreviations; such keys rust with disuse, and may fail in after times to unlock the secret that

should have been laid bare in the beginning. Use no signs intelligible only to yourself; your notebook may come to be overhauled by others whom you would not wish to disappoint. Be sparing of sentiment, a delicate thing, easily degraded to drivel: crude enthusiasm always hacks instead of hewing. Beward of literary infelicities: the written word remains after you have passed away; put down nothing for your friend's blush or your enemy's sneer, write as if a stranger were looking over your shoulder.

Even the most abundant species of mammals would repay investigation along many of the lines indicated above. Most of our shrews and bats, many of our smaller rodents, and certainly the fur-bearing predators have been sorely neglected. Many reports, some of monographic scope, have been written on our eastern mammals, yet we have only scratched the surface. During the century and more that has passed since the days when Audubon and Bachman were amassing observations for their monumental *Viviparous Quadrupeds of North America* we have added relatively little to our knowledge of many southern mammals, and of some northern species.

It is our hope that this volume will help the interested field zoologist, particularly the enthusiastic amateur, to an understanding and appreciation of our native mammals and in a small way point out neglected fields that will repay study.

Key to the Orders of Eastern Mammals

A. Female provided with an external pouch in which the young are carried for some time after birth; tail prehensile; fifty teeth, and with five upper incisors on each side; hind foot with five toes, the innermost of which is thumblike and clawless (opossums) . . . .*Marsupialia*

AA. Female without external pouch for carrying young; tail never prehensile; teeth less than fifty, and less than five upper incisors on each side; innermost toe of hind foot never thumblike.

    B. Fore limbs modified to serve as wings (bats) . . . .*Chiroptera*

    BB. Fore limbs not modified to serve as wings.

        C. Toes armed with hoofs (bison, deer, etc.) . . . .*Artiodactyla*

        CC. Toes armed with claws.

            D. Incisor teeth chisel-shaped, not more than two in lower jaw, and separated from grinding teeth by a wide space (gnawing animals).

                E. Four incisors in upper jaw, the second pair rudimentary and placed directly behind the grooved first pair (rabbits) . . . .

                . . . .*Lagomorpha*

                EE. Two incisors in upper jaw (rats, mice, squirrels, etc.). . . . . .

                . . . .*Rodentia*

            DD. Incisor teeth not chisel-shaped, not separated from the grinding teeth by a wide space, i.e., tooth row essentially continuous.

                F. Incisor teeth present.

                    G. Eyes well developed, muzzle not greatly elongated, length large, usually more than 250 mm (10 inches) (flesh eaters) . . . .*Carnivora*

                    GG. Eyes small, often concealed in fur, muzzle greatly elongated, size small, usually less than 200 mm (8 inches) . . . .*Insectivora*

                FF. Incisor teeth absent (Amercan species with bony armor). .

                . . . .*Edentata*

1

Marsupialia

(Opossums)

The marsupials are characterized by a primitive brain, epipubic bones on the pelvis, a decided inturning of the angular process of the jaw, and young born in an undeveloped state; in some species, the female possesses a prominent marsupium or pouch for the care of her young. American marsupials have numerous teeth, fifty being present in *Didelphis*.

The order has its center of abundance in Australia, but two families of marsupials occur in northern South America, and members of one of these, the Didelphidae, have reached central America. The opossum, *Didelphis virginiana*, occurs commonly over much of the United States.

Opossum. *Didelphis virginiana* Kerr

DESCRIPTION.   The opossum (also called Virginia opossum) is a marsupial about the size of a house cat, but with shorter legs, large, naked ears, and a long prehensile tail (Figure 1.1). The first toe of the hind foot is without a nail and is opposable to the others. Long, white hairs overlie the black-tipped underfur and give a grizzled appearance to the animal. The head, throat, and cheeks are whitish. The pelage is less grizzled below than dorsally. The black leathery ears are tipped with white; the tail, except at its blackish base, is generally creamish white. The female has a marsupial pouch in which the young are carried. Measurements of ten adults from Indiana, New York, North Carolina, and Pennsylvania averaged: total length, 767 (686–835) mm; tail, 321 (290–348) mm; hind foot, 68.5 (60–74) mm. Mature animals may weigh up to 9 pounds (4–5 kg) or even more. In the area from Charleston, S.C., into southern Florida, and west through the Gulf region to Louisiana, the opposum is smaller and darker, with a longer, slimmer tail, and the ears and toes tend to be all black.

Figure 1.1. Opossum, *Didelphis virginiana*. Photo by Terrie Nesslinger.

DISTRIBUTION.   Within the past century, the opossum has extended its range materially to the northward, so that it now occurs sporadically throughout Massachusetts and central New Hampshire and Vermont; its range extends northward in western New York to Lake Ontario and to central Michigan and Wisconsin, and south to the Gulf Coast (Figure 1.2).

HABITS.   The opossum is an inhabitant of open woods, swamps, and waste lands generally. Its presence is often overlooked, for the animal is shy and secretive and becomes active only at night. Although it is well adapted for arboreal life, much of its time is spent on the ground, and it may wander considerable distances from trees.

The opposum is often very abundant as indicated by numerous carcasses on the highway that have been killed by cars. Perhaps this evidence indicates the resourcefulness of the opossum in making use of a ready food source, the many other animals killed by motor vehicles. A goodly proportion of the food of the species is carrion.

The opossum makes a nest of leaves in a fallen log, a hollow tree, or a cleft in a cliff which excludes sunlight, or it may frequent a woodchuck or skunk burrow. It may pass a week or more in such well protected retreats during the severest winter weather, but occasionally it is abroad when near zero temperatures prevail. At such times the naked ears and tail may become severely frostbitten. Nest building is particularly interesting. Nest materials are grasped in the mouth, passed under the animal to the tail, which is turned forward between the hind legs. The materials are then grasped and transported by the tail.

Figure 1.2. Distribution of the opossum, *Didelphis virginiana.*

Not the least remarkable feature of this curious animal is its reproductive habits. After a remarkably short gestation of twelve to thirteen days, the tiny young, which look like abortions, are born (Figure 1.3). These embryo-like creatures crawl hand over hand from the vaginal orifice to the marsupium, where they immediately grasp a teat. Normally there are thirteen teats, although eighteen or more young may be born, with the average at about six to nine. There is no fusion of tissue, but the teat does expand inside the mouth of the young forming a relatively permanent attachment which is maintained for

Figure 1.3. Newborn opossum and honeybee.

fifty to sixty-five days. Obviously, in larger litters, those young reaching the pouch after the teats are all occupied are doomed to perish. Usually not more than seven or eight are found in the pouch a month after birth. These are about the size of grown house mice. More than two months are spent in the pouch; some solid food is taken prior to weaning, which takes place at about ninety-five to one hundred five days. When the young first clamber from the pouch they cling to the long fur of their parent. Two broods per year are normally produced, occasionally three. Pouch young are found from February through October. Science is indebted to Carl Hartman (1920) for recording many of the details of reproduction in *Didelphis*.

The opossum is omnivorous, feeding on various fruits and berries, earthworms, insects, frogs, snakes, birds, and small animals. It is said to relish particularly the persimmon, although blackberries, apples and corn "in the milk" are favorite foods. Hamilton once found a newly hatched box turtle in the stomach of an opossum, and opossums are known to eat bats. Their tendency to feed on carrion was mentioned previously, and there is little they will pass up. Of 461 New York opossums examined by Hamilton (1958), insects, mammals, green vegetation, fruits, earthworms, and amphibians constituted the bulk of the food.

The opossum has many enemies, particularly among the larger predatory mammals and owls. Its chief enemy is, of course, the speeding car. From natural enemies it may gain some measure of protection from its well-known habit of feigning death. This attempt may be quite realistic, for the animal will roll over, shut its eyes, loll the tongue from its open jaws and retain such a position for some time.

The opossum ranks among the important fur-bearers. Although the coarse fur is chiefly used for trimmings and brings the trapper much less than most of the other furs (two dollars in 1977), the great numbers taken even in such northern states as Pennsylvania and Indiana make it an important species in the fur trade. In addition, it provides much sport to the night hunter, who with his hounds bags a large proportion of those animals whose skins reach the market. Moreover, the flesh is relished by many people, and thousands are killed simply for food each year.

2

Insectivora

(Moles and Shrews)

The insectivores comprise a group of small mammals with representatives in all parts of the world except Australia and the southern two-thirds of South America. Practically all have five clawed toes on each foot, a long pointed snout extending considerably beyond the jaw, and a wedge-shaped skull with zygomatic arch lacking in the shrews and much reduced in the moles. The teeth are sharp and pointed, the canines little differentiated from the incisors or premolars (Figure 2.1). In the species of the eastern United States the eyes are minute and probably of little use. Prominent scent glands are found in most species.

Some shrews are among the smallest of mammals, certain species weighing little more than a dime. They are often numerous in the forests, marshes, and meadows, where they consume prodigious quantities of worms, insects, and other invertebrates, and in turn themselves provide food for predatory animals.

Key to Genera of Eastern Insectivores

A. Forefeet broad and enlarged, more than twice as large as hind feet, adapted for digging; body stout and cylindrical, without apparent neck . . .*Talpidae, Moles*

  B. Tail long, 60 mm or more, snout with fleshy rosette of tentacles . . . .
. . . .*Condylura*

  BB. Tail short, less than 40 mm, snout plain.

    C. Tail naked . . . .*Scalopus*

    CC. Tail densely furred . . . .*Parascalops*

AA. Forefeet not enlarged, snout long and pointed, mouselike, teeth tipped with brown . . . .*Soricidae, Shrews*

  D. Tail long, over 30 mm, ears visible.

    E. Total length more than 90 mm, four or five unicuspid teeth (Figure 2.1) visible when viewed from the side . . . .*Sorex*

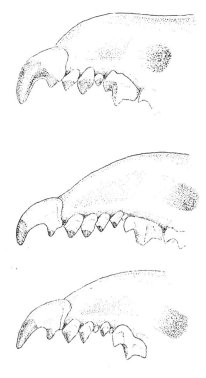

Figure 2.1. Upper partial tooth rows of some eastern shrews, illustrating disposition of the small unicuspid teeth. From top to bottom the teeth represent *Cryptotis parva, Sorex fumeus,* and *Microsorex hoyi.* All are enlarged ten times.

EE. Total length less than 90 mm, only three large unicuspid teeth (Figure 2.1) visible in profile . . . . *Microsorex*

DD. Tail short, less than 30 mm, ears hidden in fur.

    F. Length less than 90 mm, three unicuspids visible from the side . . . .
. . . . *Cryptotis*

FF. Length more than 90 mm, four unicuspids visible from the side, the fifth small and hidden behind the fourth . . . . *Blarina*

Key to Long-Tailed Shrews of the Genus Sorex

A. Color generally grayish (old *Sorex fumeus* may be brownish)

    B. Total length greater than 140 mm. Fringe of stiff hairs on edge of feet . .
. . . . *Sorex palustris*

BB. Total length less than 140 mm. No such fringe.

    C. Infraorbital foramen with posterior border lying behind space between first and second upper molariform teeth. Tail more than 50 mm . . .
. . . . *Sorex dispar*

CC. Infraorbital foramen with posterior border lying ahead of this space. Tail less than 50 mm . . . . *Sorex fumeus*

AA. Color generally brown.

    D. Distinctly tricolored, with back dark, sides lighter, and belly still lighter
. . . . *Sorex arcticus*

DD. Without tricolor pattern, brown above and light below.
  E. Tail relatively longer, generally 35–46 percent of total length, and well haired. Longest hairs at tip of tail, when unworn, about 2–3 mm long. Rostrum long and narrow. Greatest width across outside of first large molariform teeth usually more than twice the distance from posterior end of palate to anterior end of first incisors. Third unicuspid not smaller than fourth. Inner ridge of upper unicuspids with pigment (*Sorex cinereus* and *S. longirostris* are difficult to separate) . . . . . .
  . . . .*Sorex cinereus*
  EE. Tail relatively shorter, generally 32–38 percent of total length; longest hairs about 2–3 mm long. Rostrum short, wider. This width usually less than twice the distance from posterior end of palate to anterior end of first incisors. Third unicuspid usually smaller than fourth. Inner ridge of upper unicuspid lacking pigment . .*Sorex longirostris*

Masked Shrew. *Sorex cinereus* Kerr

DESCRIPTION.   *Sorex cinereus* is one of the smallest of shrews. Summer pelage is grayish brown above, pale smoke gray below. The tail is bicolor, fuscous above, buffy below. Winter pelage is darker, grayish fuscous to dark brown above, much paler below (Figure 2.2). This shrew is immediately distinguished from *Sorex fumeus* by its smaller size and browner color.

Figure 2.2. *Sorex longirostris* (right) and *Sorex cinereus* indicating greater length and amount of hair on tail and longer, thinner snout of *cinereus*.

Average measurements of fifteen adults from Long Island: total length, 92 mm; tail, 36.2 mm; hind foot, 12.1 mm. Hartley H. T. Jackson gives the measurements of three adult females from Vilas County, Wisconsin, as: total length, 101.7 (99–103) mm; tail, 38.7 (38–40) mm; hind foot, 12 mm; weight, 3–5 grams.

In the Maryland area, *S. cinereus* is smaller, with shorter tail, narrow brain case, shorter rostrum, and shorter unicuspid tooth row. The Maryland shrew was given full specific rank (*Sorex fontinalis* Hollister) by Jackson (1928), but Poole (1932) examined Pennsylvania specimens and concluded that it should stand only as a subspecies of *Sorex cinereus*. Specimens from northern Berks County, Pennsylvania, are clearly intermediate between *S. c. cinereus* and *fontinalis*. In New Jersey, winter pelage is blackish-plumbeous above and lighter plumbeous below; tail and feet dark brown.

DISTRIBUTION.   *Sorex cinereus* is one of the most widely distributed of all American mammals. In the range treated by this volume the species occurs from Maine to Wisconsin, south through northern Illinois, northern Kentucky, the Smoky Mountains of North Carolina and Maryland (Figure 2.3).

Figure 2.3. Distribution of the masked shrew, *Sorex cinereus.*

HABITS.   This tiny mammal is a true cosmopolite, ranging from the salt marshes to the high slopes of mountains above timber line. It seems as much at home in the grassy fields and marshes of the voles as in the dark moss-carpeted spruce forests. We took several along the edge of a cypress swamp in southern Indiana, but it is often very abundant in open or wooded moist habitats, particularly those carpeted with moss. Perhaps no other American mammal has such a wide range and choice of habitat.

The masked shrew is an abundant animal in many places, but appears to be relatively scarce over much of its range. Some years these little sprites swarm, and the collector will find his traps taking great numbers. At other times they are quite scarce. Quite often one's traps may take few of these in the first two or three nights of trapping, whereas the catch may increase after a few days of trapping. The reason for this is not known.

These tiny shrews are very abundant along the coast; during high water, when their burrows are flooded, large numbers may be seen on floating driftwood. They invade the trapper's cabin in the coniferous forests, and at times swarm in the northern meadows and the dense boreal woods.

A dainty nest composed of leaves and grasses is placed beneath a log or stump or in a shallow burrow. Here the incredibly small young are born from spring until early fall. Three litters appear to be the rule, and each brood may number from four to ten, a truly large family for such a tiny beast. Little is known of their home life; a careful study would surely repay the investigator.

One often finds these shrews dead in the woods, particularly in fall. It is probable that these are old animals; apparently they die after a remarkably short life, perhaps at the close of their first breeding year. The masked shrew is active at all hours, as is indicated by the times of capture, but appears most active at night.

Shrews do not hibernate. Their elfin tracks may be seen on the snow during subzero weather. It seems incredible that these tiny mammals can produce sufficient heat to maintain life under such conditions.

The food of the masked shrew is not unlike that of its relatives. Tiny molluscs, insects, small annelids, or the dead bodies of much larger animals are eaten. Their major foods in Indiana are lepidopterous larvae, coleopterous larvae, slugs, snails, and spiders.

Even during midwinter shrews find a sufficiency of dormant insects to supply their needs. Plant food is eaten sparingly, but if other more favored edibles are lacking, they may subsist on plants for many days.

Every small and abundant mammal has a host of enemies, and this tiny beast is no exception. Owls, hawks, shrikes, herons, and predatory mammals take a considerable toll. Notwithstanding the prominent lateral scent glands which these shrews possess, foxes and weasels kill and eat large numbers. Hamilton once found the sole stomach contents of a twenty-pound bobcat to be one of these shrews. Fish are known to capture them, and a specimen has been removed from a merganser.

The services of this woodland sprite, like those of other insectivores, cannot be easily measured, but it should be classed as an asset. Its aggregate destruction of insects must be considerable, and for this man is in its debt.

Southeastern Shrew. *Sorex longirostris* (Bachman)

DESCRIPTION. *S. longirostris* is similar to *Sorex cinereus* but smaller and more reddish in color. It differs further in having a relatively shorter, broader rostrum, a shorter and more crowded unicuspid row and the third upper unicuspid usually smaller than the fourth.

The tail is shorter and more sparsely haired than in *Sorex cinereus*. The head and tail proportions are illustrated in Figure 2.2.

Basal length of the skull is 14–14.2 mm. Remington Kellogg (1939) found a series of twenty specimens from Maryland previously referred to as *fontinalis,* and a similar number of *longirostris* from Virginia, Maryland, North Carolina, Georgia, Tennessee, and Indiana to be so highly variable and lacking even in limited correlation with geographic distribution that he concluded that the two were synonymous. However, examination of further specimens from Alabama, Indiana, and elsewhere has now shown them to be separate. Measurements of six adults from various localities vary as follows: total length, 78–90 mm; tail, 27–33 mm; hind foot, 10–11.5 mm. Measurements of 33 individuals from Indiana are: total length, 72–90 mm; tail, 26–33 mm; hind foot, 9–11 mm; weights, 3–4 grams.

DISTRIBUTION. *S. longirostris* occurs from southern Maryland and the District of Columbia south to northern Florida, westward through Alabama, Tennessee, western Indiana and much of Illinois (Figure 2.4).

HABITS. Like the other species of *Sorex,* this species seems to prefer moist situations. A few have been taken in bogs and damp woods, while others have been trapped on comparatively high ground, not in swamps nor the edges of them. A likely place to trap these diminutive shrews is in clumps of honeysuckle during the winter season.

In west central Indiana, where both species are taken, *Sorex cinereus* is generally taken in the bottomlands, *S. longirostris* in the uplands, although occasionally they are taken together. A good way to trap this and many other shrews is by using cans sunk to ground level.

Little is known about the habits of this shrew, although it is not extremely rare as once thought. Its habits seem quite similar to those of *Sorex cinereus*. It feeds on various invertebrate material. Seven Indiana individuals had eaten

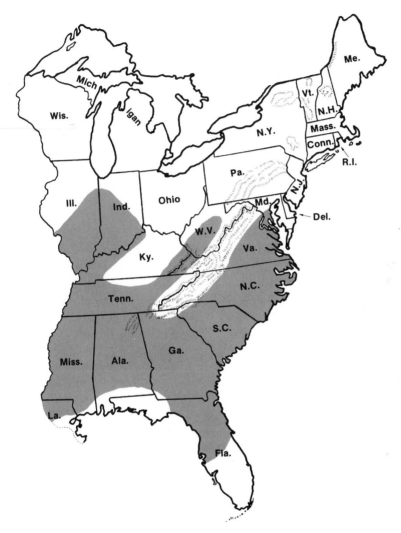

Figure 2.4. Distribution of the southeastern shrew, *Sorex longirostris.*

primarily spiders, lepidopterous larvae, slugs and snails, vegetation, and centipedes.

Like other shrews, it has a host of enemies. The stomach of an opossum
produced one in Virginia; South Carolina specimens were recovered from a
barn owl and from the throat of a hooded merganser shot in a rice marsh, and
the single specimen from Alabama had first been collected by a barred owl.

William L. Engels (1941) found four young in a nest near Chapel Hill,

North Carolina, on April 18, 1940. The young weighed 2.5 grams each and averaged 71 mm in total length. The nest was simply a shallow depression lined with fragments of leaf litter, placed beneath a well-decayed log in an oak-pine forest.

Water Shrew. *Sorex palustris* (Richardson)

DESCRIPTION.   The water shrew is a large, soft-furred, dark-colored shrew partially modified for an aquatic life (Figure 2.5). The large broad hind feet are conspicuously fringed with stiff hairs, while the toes are slightly webbed at their base. In winter pelage, the upper parts are dark fuscous black; they are slightly paler and with more brown in summer. The underparts are grayish white. The tail is dark brown to blackish above, only slightly lighter below at the tip but distinctly bicolored at the base (Figure 2.5). Individuals from Wisconsin and northern Michigan are similar, although markedly different in color of winter pelage, when they are fuscous-black to blackish mouse-gray above; below, pale smoke gray to pale olive gray, touched over all with silver; the color of ventral surface extends onto the upper lips and part of the flanks; the tail is bicolor, blackish brown above, much paler below. Measurements of six adults from New York and Maine averaged: total length, 140 mm; tail, 66 mm; hind foot, 19 mm; weight, 12–17 grams. Individuals

Figure 2.5.   Water shrew, *Sorex palustris.*

from the southern Allegheny Mountains are larger, have grizzled upper parts, pale gray underparts in winter pelage, and a narrow interorbital space.

DISTRIBUTION. Water shrews are found in the northeastern United States, from Maine to Connecticut, westward to southeastern Pennsylvania and eastern (Westchester County) and northern New York; in Wisconsin and the upper peninsula of Michigan, and in the southern Allegheny Mountains (Figure 2.6).

HABITS.   As their name implies, water shrews frequent wet places, often occupying the shoreline of rushing mountain streams or the sphagnous swamps bordering beaver meadows. We have collected several on the slopes of Mount Katahdin, Maine, along the sandy shelf of mountain brooks, the mud flats of sluggish backwaters, and the mosscovered boulders of yellow birch and striped maple thickets.

These little mammals swim and dive with great celerity, and can actually run on the surface of the water with facility. H. H. T. Jackson (1961) observed a water shrew near Rhinelander, Wisconsin, run a distance of about five feet across a small pool, the surface of which was glassy smooth. According to Jackson, the body and head of the animal were entirely out of water, the surface tension of the water supporting the shrew, and at each step the animal took there appeared to be held by the fibrillae on the foot a little globule of air, which was also discernible in the shadow at the bottom of the pool.

A captive water shrew (*Sorex p. navigator*) observed by Arthur Svihla used all four feet when swimming. It had difficulty in forcing its way to the bottom because of the air entrapped in its hair, which lent much buoyancy. On reaching the bottom it "literally stood on its long flexible nose which was thrust into the sand and debris searching for food, its feet kicking rapidly in order to maintain this position. To come to the surface again it merely stopped kicking and immediately rose like a cork. The length of time it remained submerged varied from a few seconds up to a quarter of a minute—on coming to the surface a few shakes of the body make the coat dry."

The stomachs of thirteen contained various insects (among them the aquatic nymphs of stoneflies and mayflies), planarians, and a little plant material. A friend was once troubled with some small predator which continually pilfered the trout eggs from trays sunk in flowing water. Baiting a weighted mouse trap with these eggs, he was rewarded by catching several water shrews.

Little is known of the reproductive habits of this shrew. One taken in North Carolina on April 20 contained five embryos, and specimens have been collected containing seven embryos. It is probable that, as with other members of the genus, two or three litters are produced in a season.

Although the species is not often collected, it is far more abundant than

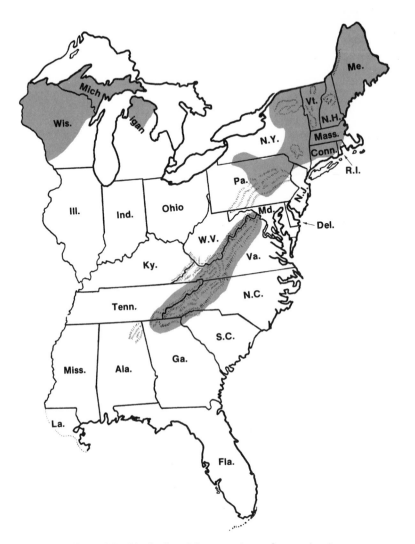

Figure 2.6. Distribution of the water shrew, *Sorex palustris.*

museum collections would indicate. A specimen was removed from the stomach of a trout that was caught twenty miles north of New York City. One fortunate enough to live in a region where the species is common could make an interesting and worthwhile contribution to mammalogy by a close study of this handsome big shrew.

Smoky Shrew. *Sorex fumeus* (Miller)

DESCRIPTION.   Aside from *S. palustris* and *S. dispar,* both rather un-common in the east, *Sorex fumeus* is the largest of the eastern long-tailed shrews. The colors are markedly different in winter and summer. Winter pelage is dark mouse gray above and rather lighter below. In summer the fur is much browner, approaching olive-brown and paler below, occasionally sil-very on the belly. The tail is bicolor, fuscous above and paler below. The ears are relatively prominent (Figure 2.7).

Average measurements of twenty-five adults of both sexes from Ithaca, New York, are: total length, 116.8 (110–125) mm; tail, 42.6 (37–47) mm; hind foot, 12.8 (12.3–14) mm. Weight, 7.7 grams (6.1–11). The young may be distinguished from adults by their smaller size, seldom exceeding 6.5 grams, and by the pencil of long hairs on the tail tip. In the adult the naked tail tip is rounded and smooth.

DISTRIBUTION.   The smoky shrew occurs in eastern North America from Nova Scotia to the Smoky Mountains and northern Georgia, westward to Wisconsin. Although this shrew has not been taken in Illinois, Indiana, or Michigan, Vernon Bailey collected a small series in Mammoth Cave, Ken-tucky (Figure 2.8).

HABITS.   Shady damp woods, either hardwoods or conifers, are the choice of these shrews. Extensive clumps of yew, and moss-covered logs and boulders in maple, birch, and hemlock woods usually harbor a number. Unlike

Figure 2.7.   Head of smoky shrew, *Sorex fumeus.*

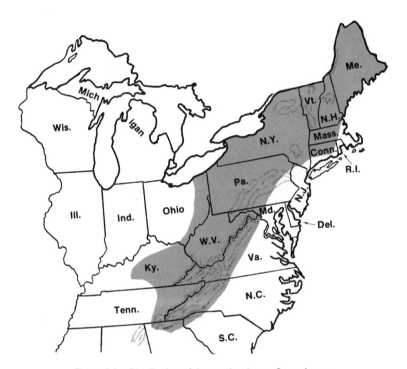

Figure 2.8. Distribution of the smoky shrew, *Sorex fumeus.*

the smaller *S. cinereus,* the smoky shrew is seldom found in dry woods and never in the marshes bordering the coast.

Traps set at the entrances to tiny burrows leading into a bank along shaded brooks in New York and along mountain roads in North Carolina have often yielded smoky shrews.

In the damp, dark northern woods which support a dense growth of ferns and other ground cover, shaded by a canopy of second growth timber, a seldom-seen fauna of small mammals exists. The smoky shrew has for company the deer mice, red-backed mice, jumping mice, Blarinas and hairy-tailed moles, and travels the burrows of these other forest sprites, or threads through the leaf mold of the forest floor. In some seasons this species is not uncommon and several will be caught in a dozen traps set for one or two nights. At other times it is very scarce. The smoky shrew is not so generally distributed throughout its habitat as are deer mice and *Blarina*; it seems to colonize to some extent.

The favorite haunts of this species appear to be the well-traveled highways deep in the leaf mold which covers to several inches the roots of venerable hemlocks and birches. A nest of shredded leaves or grasses is placed in the honeycombed recesses of some half-rotted log or moss-covered stump or directly beneath a slab of rock.

During the spring the tails of these shrews become swollen, a phenomenon much like that of *Condylura*. The first litter is born in late April. The young may number from four to seven and are born blind and naked. Before another month has passed a second litter is on its way. Hamilton (1940) has collected gravid females in early July and, rarely, in August. Thus two litters are customary, and less often a third litter is produced in the fall.

It is seldom that one collects an adult shrew in the late fall or winter. This makes it seem likely that the adults die off in their second year, following the breeding season.

Like other shrews, this species does not hibernate. This fact was particularly well emphasized one winter when Whitaker set about two hundred mousetraps in a forest at Oneonta, in central New York, during a very cold period. For five days in a row morning temperatures were not above minus twenty degrees, and noon temperatures were not above minus five. On the morning in question, the temperature at 8:00 A.M. was minus thirty-one degrees Fahrenheit ($-30.5°$ Celsius) but three *Sorex fumeus* were taken.

The principal food of this little insectivore is the invertebrate population of the leaf mold. Insects and their larvae, centipedes, small salamanders, plant matter, earthworms, sowbugs, and other small fry make up the greater share. The shrew's protruding, pincerlike incisors are efficient organs with which to gather each tiny morsel that it comes upon.

We may number among the enemies of the smoky shrew all the owls and some of the hawks, foxes, weasels, and other mammalian predators, and *Blarina* sometimes prey upon the smaller shrews which inhabits its burrows.

Whitaker once took two shrews of this species together in the same snap trap in an Ithaca, New York, swamp. One was a pregnant female, the other was a male. The ground was of soft dirt covered with moss and was torn up in a circle about ten inches in diameter. The male was firmly grasping the fur of the flank of the female and had penetrated the skin slightly. It appeared that the two shrews had been in pitched battle when killed by the trap.

These tiny mammals and their confreres are important checks on the undue increase of many insects. Their effectiveness in destroying the larvae and pupae of important forest insect pests can hardly be denied.

Arctic Shrew. *Sorex arcticus* Kerr

DESCRIPTION.   Compared with other eastern members of the genus, the arctic shrew or "saddle-backed shrew" may at once be recognized by the peculiar tricolored pattern of the pelage: the back is distinctly darker than the sides, which in turn are darker than the belly. In winter the back is fuscous black or blackish brown; sides cinnamon brown; belly gray, tinged with buff; tail colored as back, slightly paler below. The summer pelage is much paler, usually cinnamon brown above, the transition to paler brown on sides less marked than in winter specimens; the belly darker than in winter, with more brown. The average measurements of three adults from Wisconsin are: total length, 116 mm; tail, 40 mm; hind foot, 14 mm; weight, 6–9 grams (Figure 2.9)

DISTRIBUTION.   In the eastern United States, the arctic shrew occurs in Wisconsin and the upper peninsula of Michigan (Figure 2.10).

HABITS.   Little is known of the habits of this handsome fellow. Bernard Bailey collected Minnesota specimens from the marshes surrounding rice lakes and among old stumps in tamarack swamps. F. J. W. Schmidt (1931) collected a specimen in Clark County, Wisconsin, in a similar habitat.

Specimens which Bailey took in Minnesota on April 27 and June 1 contained six and nine embryos respectively, and Jackson (1961) reported four females with from six to nine embryos each. From this it would appear that several litters are produced each season.

An interesting situation arose concerning myobiid mites, *Protomyobia brevisetosa*, which Whitaker and Pascal (1971) reported from this shrew. Letters were soon received from both E. W. Jameson, Jr., and F. Dusbabek asking if it was not *P. onoi*, which they had just described from *Sorex araneus* and *S. unguiculatus* from Eurasia, and which they had predicted on geographical and ecological grounds might be on *Sorex arcticus*. They were

Figure 2.9. Arctic shrew, *Sorex arcticus.*

Figure 2.10. Distribution of the arctic shrew, *Sorex arcticus.*

correct. This episode is related primarily to indicate the mostly untapped use host-parasite relations can sometimes play in mammalian taxonomy and zoogeography.

Long-Tailed Shrew. *Sorex dispar* Batchelder

DESCRIPTION.   The long-tailed shrew is similar in appearance to *Sorex fumeus,* but may be recognized by the longer tail, slimmer body and nearly uniform darker coloration, the belly being almost the same color as the back. Summer pelage is dark mouse gray above, belly scarcely paler; tail fuscous black above, yellowish brown below, brighter near its base. In winter the fur is softer and slate-colored, nearly uniform throughout. Average measurements of thirteen specimens from New York are: total length, 121.6 mm; tail, 56 mm; hind foot, 14.5 mm; weight, 5–6 grams.

DISTRIBUTION.   The long-tailed shrew is not well represented in collections, there being only two dozen or so locality records from Maine through the mountainous areas to North Carolina and Tennessee (Figure 2.11).

HABITS.   This slender, dark-colored little shrew favors moist rocks, crevasses between boulders, and the moss-covered logs of damp coniferous forests. Several have been secured in the recesses of large masses of boulders, where the shade permits the ice to linger into July, and indeed, the best way to trap them appears to be to set mousetraps deep in the dark recesses of talus slopes. Paul C. Connor (1960) has obtained several specimens in such a manner at Gilboa, Schoharie County, New York.

To the inexperienced collector, this species might pass for the smoky shrew, save for its long tail and almost uniform slate color. This pelage persists into the summer and is then strikingly different from that of *Sorex fumeus.*

Little is known of its habits, but it would seem likely that some differences

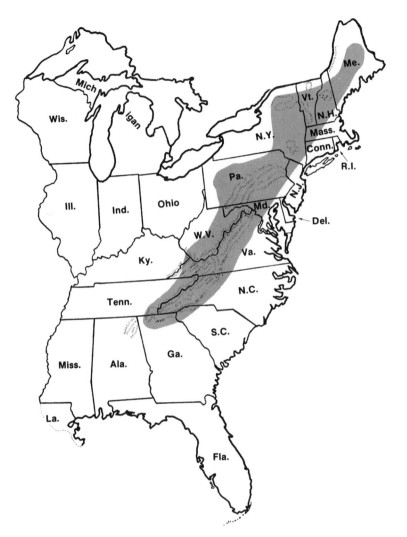

Figure 2.11. Distribution of the long-tailed shrew, *Sorex dispar.*

exist relating to its habitat in the recesses beneath boulders. Connor found the major foods of nine from Schoharie County, New York, to be adult Diptera, *Ceuthophilus* (Orthoptera), spiders, centipedes, and Coleoptera.

Morris M. Green (1930), who captured one of these shrews on North Mountain, Pennsylvania, compared it to a swift, lightweight cruiser and the smoky shrew to a heavier, slower battleship. While Green appeared doubtful that the former can exist in company with the latter, we found them together on Mount Whiteface, in the Adirondacks of New York, and Earl Poole found *Sorex cinereus, S. fumeus,* and *S. dispar* all in the same locality. Poole suggests the possibility of a greater difference in habits than is at present known.

Pygmy Shrew. *Microsorex hoyi* (Baird)

DESCRIPTION.   In general appearance the pygmy shrews closely resemble *Sorex cinereus.* They differ by their slighter build and shorter tail, but they can be determined definitely only by observing the teeth. The third unicuspid in the upper jaw is very small, wedged in between the second and fourth unicuspid (Figure 2.1): it is so exceedingly small that magnification is necessary to see it, and even then it cannot be observed from the outside. Moreover, the fifth unicuspid is also tiny, so only three unicuspids usually appear on the outside, whereas in *Sorex cinereus* five may be seen. The color not unlike that of *cinereus,* sepia brown above, smoke gray below, tinged with light buff. In winter, the general color of upperparts is dark brown to olive-brown; smoky gray below, occasionally tinged with buffy; tail indistinctly bicolor at all seasons, dark brown above, darker at tip (Figure 2.12). Average measurements of four specimens from Minnesota are: total length, 82.2 mm; tail, 30.7 mm; hind foot, 10 mm; weight, 2.5-4.5 grams. Average measurements of thirteen Maine specimens are: total length, 85 mm; tail, 29.5 mm; hind foot, 9.4 mm; weight, 2.5-3 grams. Specimens from the southern Alleghenies

Figure 2.12. Pygmy shrew, *Microsorex hoyi.*

are the smallest of the genus and possibly constitute the *smallest mammal in the world*. The type, from Stubblefield Falls, Fairfax County, Virginia, measures: total length, 78 mm; tail, 28 mm; hind foot, 9 mm. A specimen from the Pisgah Forest of North Carolina measured: total length, 80 mm; tail, 28 mm; hind foot, 10 mm. It weighed only 2.3 grams, or somewhat less than a dime.

Charles A. Long (1972) proposed separating *M. h. thompsoni* and *M. h. winnemana* as a separate species, *M. thompsoni,* on the basis that "where *M. hoyi* approaches *M. thompsoni* (in the Great Lakes region) the former is decidedly larger, less grayish, has a flattened cranium (ratio of cranial breadth to depth usually less than 1.7 in old adults), and possesses larger teeth." Long goes on to state that this view needs further study in critical areas. We do not see sufficient basis to establish a new species, and thus have retained eastern *Microsorex* in a single species here.

DISTRIBUTION.   From Maine westward through New York to Wisconsin, and in the eastern mountains, from Maryland and Virginia to North Carolina (Figure 2.13).

HABITS.   Pygmy shrews are quite rare in collections, and their habits are little known. Their rarity may be apparent rather than real, at least in some places. This is attested by W. E. Saunders (1932), who writes of his collecting experience in Ontario thus:

"The vital necessity of a magnifying glass may be well authenticated by the following incident. At Noganosh, in 1930, we were finding *Sorex cinereus* common—too common, in fact, and we threw away quite a number. One day Mr. Davis was making one up, and some trivial incident prevented him from finishing the skin, and it was thrown away. What was his disgust, on reaching home and examining the skull, to find that the specimen was not *cinereus* at all, but one of this species, so excessively difficult to procure. We were going into ecstasies over the capture, every second day, of an additional water shrew, but the pygmy shrew, *Microsorex hoyi,* is infinitely rare to most collectors. The prompt use of a glass would have prevented all the chagrin and would have saved this rare specimen."

The shrews make tiny burrows beneath stumps, fallen logs, and the leaf carpet of the forest. We have taken specimens in open fields along with *Sorex cinereus* at Ithaca, New York, and along a woods edge at Oneonta, New York. Paul Connor has taken several in woods. One which F. J. W. Schmidt (1931) observed in Wisconsin was eating the contents of a dung beetle burrow, being able to enter the burrow without enlarging it to any extent. Another which P. B. Saunders (1929) captured at Clinton, New York, made holes in the dirt floor of his cage which could easily have been mistaken for those of a large earthworm.

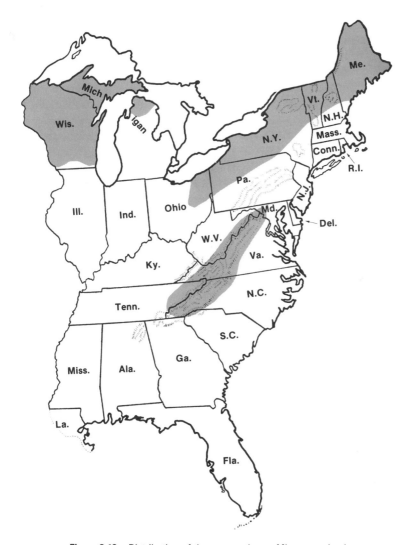

Figure 2.13. Distribution of the pygmy shrew, *Microsorex hoyi.*

Apparently the scent glands, common to all shrews, are better developed in these little animals than others of the family. Several investigators have remarked on the pronounced effluvium and the well developed side glands.

The few breeding records suggest that these tiny creatures bear litters during the summer, the usual number of young being five to eight. Long suggests that in Wisconsin, at least, the species may bear only one litter per year and that this might account for the low abundance of the species.

Not much is known regarding their food habits, although there is little reason to suspect these differ in any essential manner from those of other small shrews. The stomach of a Vermont specimen contained several fragmentary small grubs, a segment or two from a small earthworm, and several chitinous parts from an undetermined insect, quite probably a beetle. Caterpillars, beetle larvae and adults, adult flies, and other insects have been found in stomachs.

The collector should carefully examine every small *Sorex,* in expectancy of capturing *Microsorex.*

Key to Short-Tailed Shrews of the Genus Blarina

Blarina brevicauda and *B. carolinensis* were long recognized as belonging to the same species, but recently they have been found occurring together in southeastern Nebraska, without interbreeding. More work is needed to determine their relationships in the southeastern United States, but we have tentatively classed them as separate species. *Blarina brevicauda* is much larger and is more northern than *B. carolinensis.*

A. Hind foot (including claw) 13 mm or greater; condylobasal length of skull 20 mm or greater . . . .*B. brevicauda*
AA. Hind foot less than 13 mm; condylobasal length less than 20 mm . . .*B. carolinensis*

Short-Tailed Shrew. *Blarina brevicauda* (Say)

DESCRIPTION. *Blarina brevicauda* is a large, bobtailed, slate-colored shrew. The short legs, minute eyes, and concealed ears are good field characteristics. The sharp-pointed muzzle, which is relatively blunt in old individuals, is not nearly so pointed as in the long-tailed shrews. All American shrews have deep reddish-brown or chestnut-tipped teeth, this color being less noticeable as the tooth wears down with advancing age. Adults are slaty above and paler beneath; new fur is glossy; the short tail and feet are dark gray above and paler below. Young individuals are much darker than the adults and sometimes appear almost black. Individuals from the Dismal Swamp of Virginia are more plumbeous, with relatively longer hind feet, a narrower skull and peculiar molariform teeth. The color is dark plumbeous or slate gray above and below, but slightly darker on the rump and nose; the feet and tail are blackish.

The average measurements of thirteen specimens from Lake Drummond, Dismal Swamp, are: total length, 119.5 mm; tail, 26.4 mm; hind foot, 16 mm. Short-tailed shrews from Fort Myers, Florida, are similar to the large northern *Blarina,* but geographically separated. The Fort Myers population is surrounded by populations of *Blarina carolinensis,* from which it differs in being larger and darker. Also, the adult winter pelage is darker than in other races of *brevicauda.* Average and range of measurements of twenty-seven Florida individuals were: total length, 109 (100–116) mm; tail, 23.5 (22–25) mm; hind foot 14.1 (13.5–15) mm; weight 13.8 (11.1–17.0 grams).

Average measurements of sixty adults of both sexes from central New York are: total length, 124 mm; tail, 24.8 mm; hind foot, 15 mm. The weights of fifty adults of both sexes average 19.3 grams. Large individuals may weigh 27 grams, while juveniles weigh but 12–15 grams (Figure 2.14).

DISTRIBUTION.   *Blarina brevicauda* is found in all the eastern United States south to Kentucky, and in the mountains to Georgia and Alabama. It is also known from Lee County, Florida (Figure 2.15).

HABITS.   The ubiquitous *Blarina* is found in a wide variety of habitats, from the salt marshes of the coastal areas to stunted timber growth on the mountains. It is usually most abundant in damp woods which support a thick leaf mold, but is not uncommon in wet meadows and overgrown fields.

The powerful little paws and stout cartilaginous nose of *Blarina* are a real asset as it plows through the leaf mold and forest litter or burrows in the loose damp soil. So common is it in such situations at times that the mammal collector despairs of securing more valuable specimens. Not only is it caught in the traps placed for rarer forms, but it falls upon the unlucky victim which has been trapped and leaves nothing but a neatly turned-back skin and a few bones.

Some years the woods fairly swarm with these shrews, while at other times they are quite scarce.

The short-tailed shrew constructs a bulky nest of partly shredded leaves and grasses, placed beneath some fallen log or stump. In this the young are

Figure 2.14.   Short-tailed shrew, *Blarina brevicauda.*

Figure 2.15. Distribution of the short-tailed shrew, *Blarina brevicauda.*

born from early spring until late September. These usually number five to seven, but as many as nine have been recorded. Three and perhaps four litters are produced each year. Young shrews grow amazingly fast and are half grown when they are a month old, accounting for the fact that nearly all shrews seem to be full grown when trapped.

In spite of their reproductive ability, it is probable that they, unlike many mice, do not breed during the season in which they are born.

Most shrews feed principally upon the teeming invertebrate fauna which they find in the soil. Insects, spiders, centipedes, snails, slugs, and to a lesser extent small salamanders, mice, and an occasional bird, are destroyed. However, earthworms are by far the single most important food of this species, followed by slugs and snails, lepidopterous larvae, and other insects. The short-tailed shrew practices at least limited food storage. Earthworms and snails are paralyzed by being bitten and then stored temporarily in piles under logs or in burrows. The species is one of very few mammals that produces a toxic secretion in the salivary glands. This secretion is apparently useful in paralyzing the earthworms and snails and perhaps in subduing other prey. It has often been said that these shrews frequently feed on small rodents, but stomach analysis has shown this not to be true. Vegetable food is not disdained, for they are known to feed on roots, beechnuts, fruits, and berries. The subterranean fungus, *Endogone,* is often eaten by *Blarina.*

Owing to their numbers, shrews have many enemies, but not a few predators discard them because of their odor. The odor arises from a pair of glands located on the sides near the flanks, and is particularly evil-smelling during the season of reproduction. Often one is aware that a shrew has been taken before it is seen in a trap, as even the human nose can pick up this odor. Hawks, owls, weasels, skunks, snakes, fish, and other enemies kill great numbers. We once found in the stomach of a red fox three shrews of this species which had been bolted entire.

Blarina has a well developed repertoire of squeaks and clicks, including ultrasonic sounds used in echolocation.

These shrews are beneficial little creatures. They do no damage to man's crops, but are of great service in ridding the fields and woods of an overabundant insect population. It is said that in New Brunswick they once destroyed 60 percent of the larch sawfly population.

Southern Short-Tailed Shrew. *Blarina carolinensis* (Bachman)

DESCRIPTION.   This species is a small edition of *Blarina brevicauda,* the total length seldom exceeding 100 mm. The skull is less massive; the fifth unicuspid is usually not visible from the outside. This species occurs in the austroriparian fauna from Virginia west to Arkansas and south to Mississippi and Florida (Figure 2.16). In peninsular Florida it differs in being uniformly slate black above and duller below, lacking the sepia-brown tint of its more northern relatives. In addition, it has a larger hind foot and more massive skull and teeth.

DISTRIBUTION.   This shrew is found from Virginia to southern Illinois southward. It is unknown in the southern Alleghenies and northwestern Georgia.

Figure 2.16. Distribution of the southern short-tailed shrew, *Blarina carolinensis*.

HABITS.   The habits of this species have not been well studied, although they are presumably similar to those of *Blarina brevicauda*.

Least Shrew. *Cryptotis parva* (Say)

DESCRIPTION.   A small edition of *Blarina*, only slightly smaller than *Blarina carolinensis*, from which it may immediately be distinguished by examining the dentition, which numbers thirty teeth, in contrast to the thirty-two teeth of *Blarina*. In winter it is brownish gray to slate, silvery gray below; the slender tail is similarly colored. Measurements of ten New York and Virginia adults average: total length, 83 mm; tail, 17.5 mm; hind foot, 11 mm. Twenty-five specimens from Raleigh, North Carolina, average: total length, 75 mm; tail, 16.4 mm; hind foot, 10.6 mm; weight, 4.5–5 grams (Figure 2.17).

DISTRIBUTION.   *Cryptotis parva* is found from New York south to Florida, westward to Illinois and Louisiana (Figure 2.18).

Individuals from Florida and southeast Georgia are similar in appearance to typical *parva,* but are somewhat darker and with longer tail. Winter coloration is plumbeous above, mixed with paler hairs which lend a peppery appearance; silvery below; some individuals have a yellowish brown patch in front of forelegs. Summer pelage is paler, inclining to brown or sepia. An adult male from Fort Myers, Florida, collected on March 25, 1940, was molting. There was new fur on the top of the head and the small of the back. Measurements of five adults from southeastern Georgia and Florida average: total length 79.5 mm; tail, 21 mm; hind foot, 11 mm; weight 4–5 grams.

HABITS.   This diminutive shrew has been trapped on the marshes about New York City, New Jersey, and Virginia, in dry fallow fields and stubble throughout the northern part of its range, and less often in meadows, in contrast to its southern relatives.

In Florida it lives along the lake shore in the dense hammocks of the prostrate cabbage palm or in the marshy woods, or travels the twisting runways of the cotton rats in old fields grown to weeds and sedges. We have caught several in traps that were placed in *Sigmodon* runways.

Figure 2.17.   Least shrew, *Cryptotis parva.*

Figure 2.18. Distribution of the least shrew, *Cryptotis parva.*

The life history of the least shrew is not well known. It seldom falls to the traps of the collector, but must be much commoner than is supposed, for owl pellets often contain a surprisingly large number in the very region where the mammalogist has failed to take it.

The nest is a ball of shredded leaves and grasses under a rock slab, stump, or log or in a shallow tunnel. Hamilton has found two nests containing five and three occupants, all adults, and H. H. T. Jackson (1961) found twenty-five individuals in one nest in Virginia. These data argue for their tolerance of one another, an attitude that is not generally accredited to shrews. A globular nest found near Sanford, Florida, by Clarence F. Smith, was loosely constructed of leaves of the panic grass and was about five inches in diameter. It was situated under one end of a prostrate cabbage palm. Several tunnels led from the nest, which was occupied by two shrews.

Several litters of young numbering from four to six are born from March to October. A female and her nearly grown litter of six young were taken from a nest in eastern Kentucky during mid-April by W. Welter and D. Sollberger. The combined weight of the litter was 17.5 grams, while the parent weighed but 5.4 grams. The mother first killed the insects which the young fed upon. Before they were fully weaned, the young shrews nursed from a parent weighing one-third as much as the combined youngsters. Hamilton captured a female from a salt marsh at Chincoteague, Virginia, that gave birth to six tiny naked young, the combined weight of these not exceeding 2 grams. The young are not weaned until nearly three weeks old.

Various small insects, molluscs, spiders, and earthworms are eaten, also the dead bodies of small mammals and other little fry. Major foods in 109 Indiana specimens in decreasing order of use were lepidopterous larvae, coleopterous larvae, slugs and snails, spiders and crickets (Whitaker and Mumford 1972).

Stewart Springer records four of these shrews huddled together under brush at Englewood, Florida. Two of these lived amicably together for several days. One, which lived more than a month, consumed quantities of mealworms, in addition to wax moths, ant lions, cockroaches, crickets and other insects, tree frogs, small-mouth toads, lizards, and a nine-inch snake (*Tantilla*). The shrew was seen to drink frequently. This shrew, whose initial weight was 4.97 grams, consumed no less than 3.5 grams of food every twenty-four hour period, with a maximum of 8.1 grams and an average of 5.5 grams. Thus the shrew consumed considerably more food than its own weight each day. This little shrew uttered a noise which Springer likened to the call of a flicker heard at a considerable distance. It was audible for only about twenty inches.

This little mammal has been called the "bee shrew" because it occasionally enters beehives and there builds its nest, feeding upon the bees and their larvae.

Owls are probably the chief enemies of this shrew. W. B. Davis (1938) reports the remains of 171 *Cryptotis* in the pellets of barn owls in Texas. These comprised 41 percent of the total number of mammals taken, and there are many other simiiar later reports. These records indicate not only the abundance of these shrews at times, but the importance of the owl as a predator. Hawks, snakes, and predatory mammals, including dogs and cats, capture these diminutive mammals as the opportunity affords.

Hairy-Tailed Mole. *Parascalops breweri* (Bachman)

DESCRIPTION.   The hairy-tailed mole is our only eastern mole with a short hairy tail. The color is dark slate to fuscous black, slightly paler below. The small eyes are well hidden in the fur. The large palms are nearly circular; the toes are not webbed. The soft thick fur is slightly coarser than that of *Scalopus*. Average measurements of twenty adults from New Hampshire, New York, and Pennsylvania are: total length, 158 (150–170) mm; tail, 27.5 (24–30) mm; hind foot, 19 (17–21) mm. Males are slightly larger than females. The weight varies from 40 to 64 grams (Figure 2.19).

DISTRIBUTION.   The hairy-tailed mole ranges through northeastern United States and adjacent Canada. It occurs from southern New Brunswick and Quebec south to the mountains of Virginia, West Virginia, and North Carolina, and occurs west to western Ohio (Figure 2.20).

HABITS.   These moles occur from near sea-level in the northern part of the range to altitudes of 3,000 feet or more in the Appalachians. They may be found in regions ranging from pasture lands that support shrubs to well wooded forests of birch, hemlock, and pines. *Parascalops* prefers a well drained, light soil and is seldom taken in damp areas or clay soils.

The hairy-tailed mole is undoubtedly more abundant than trapping records indicate. During the breeding season, W. Robert Eadie (1939) took eleven specimens from an acre in New Hampshire, but this high yield probably represents an influx into an unusually suitable habitat.

Figure 2.19.   Hairy-tailed mole, *Parascalops breweri*.

Figure 2.20. Distribution of the hairy-tailed mole, *Parascalops breweri.*

The irregular and abundant subsurface runways may form a complete network of highways, but the deeper, permanent tunnels are fewer. As winter approaches, these deeper tunnels are repaired or new ones made, and with the advent of freezing temperatures the mole deserts the upper strata of tunnels and occupies these deeper burrows. Nests about six inches in circumference are constructed of dead leaves and are placed ten to twenty inches below the surface.

The hairy-tailed mole often wanders from its tunnels during the night to feed on the forest floor. This habit is reflected in the numbers which are caught by cats and other nocturnal predators.

Mating occurs in late March or early April. The four or five young are produced in late April or May after a gestation period of probably a month. The young are blind and helpless at birth but, like other moles, grow rapidly and are probably weaned and able to shift for themselves when about four weeks old. They are sexually mature the following spring.

According to Eadie this species feeds chiefly upon earthworms, insects and their larvae, and other arthropods. Those which Hamilton has examined contained coleopterous larvae and ants. A captive mole which weighed 50 grams devoured 66 grams of earthworms and insect larvae in twenty-four hours. Moles may store earthworms after paralysis is induced by biting them.

In mountain resorts, this species sometimes does extensive damage to golf greens; but on the whole it seldom interferes with man.

Eastern Mole. *Scalopus aquaticus* (Linnaeus)

DESCRIPTION.   Moles of the genus *Scalopus* are characterized by a short, nearly naked tail and heavy broad palms. The soft velvety fur dorsally is fuscous to blackish brown in winter, paler in summer; the underparts are grayish (Figure 2.21). Measurements of a series of males from different parts of the range give the following extremes: total length, 152–184 mm; tail, 22–30 mm; hind foot, 18–21 mm. A series of females from different localities give extremes of: total length, 144–165 mm; tail, 15–28 mm; hind foot, 18–21 mm; weight, 40–50 grams. Measurements of twelve adults from northeastern Florida average: total length, 142 mm; tail, 24.5 mm; hind foot, 17.8 mm. In Hillsboro and Pasco Counties, Florida, *Scalopus* is still smaller. The total length usually does not exceed 140 mm, nor the hind foot 17 mm, in the region north of Tampa Bay. Midwestern moles are dark brown above in winter, grayish below; grayer in summer. Length of males varies from 188 to 210 mm; tail, 30–40 mm; hind foot, 21–24 mm. Females are smaller; weight 80–90 grams.

Figure 2.21.   Eastern mole, *Scalopus aquaticus.*

DISTRIBUTION. *Scalopus* is found from southern Massachusetts through Connecticut westward to Illinois, south to Florida and Louisiana (Figure 2.22).

HABITS. The eastern mole prefers a well-drained loose soil. It usually frequents open fields and pastures, but also occurs in thin woods and meadows. Following a rain, new superficial burrows, just below the surface of the ground and often marked by ridges, are pushed in all directions to

Figure 2.22. Distribution of the eastern mole, *Scalopus aquaticus.*

facilitate the capture of worms and other soil life. Ten inches or more below the surface the regular permanent highway is constructed, and the mole retreats here during long periods of dry weather or when frost is in the ground. When making fresh burrows, the mole will push excess soil up through vertical shafts, thus producing "mole hills." During burrowing, or probably at any other time, moles use the nose as a tactile organ, constantly poking it here and there. In friable soil, moles can burrow at a rate up to eighteen feet per hour. Moles do not hibernate and are active throughout the winter.

The nest is of leaves and grasses and is constructed from several inches to a foot or more below the surface. It is usually placed beneath a boulder, stump, or bush and has several approaches, one of which usually enters from below. One nest is used in winter, several in summer.

The eastern mole is much more abundant in our southern states than it is in the north. Cultivated fields will often be riddled with their burrows after a soaking rain. Moles are active at all hours, with peaks near dawn and dusk. Individuals often live two and sometimes even three years.

This mole, in common with other species, has a single litter of two to five young in the early spring, after a gestation of about forty-five days. A nest containing four young was found in mid-April on Long Island, but in the south Atlantic states the young are born earlier, perhaps in March. They are blind and naked at birth, but are exceptionally large for the size of the mother. When ten days old, they have a fine velvetlike covering of light gray fur which is retained for several weeks. As with other talpids, growth is rapid and the young are able to leave the nest and shift for themselves when about four weeks old.

The principal food is earthworms, although larvae and adult insects are under some circumstances the most prominent fare, and many other foods, such as slugs, snails, centipedes, and ants are eaten. It has been conclusively demonstrated that moles under normal conditions will eat vegetable matter, and it seems not improbable that they may take considerable quantities at times. We have seen Indiana specimens whose stomachs were completely filled with grass seeds (Whitaker and Schmeltz 1974).

The single annual litter suggests that moles have few enemies. Their fossorial habits prevent hawks and owls from taking any considerable numbers in spite of the fact that they are active at all hours. Nevertheless they do fall prey to predatory birds and mammals. Snakes can and probably do enter their burrows to seek the moles.

Moles are harmful when they disfigure lawns and provide highways in gardens for field and pine mice. Their destruction of insects places them in a more favorable light. A friend once said that these moles had almost eliminated the larvae of Japanese beetles on his grounds. Tunneling activities of moles aid in the formation of soil.

Star-Nosed Mole. *Condylura cristata* (Linnaeus)

DESCRIPTION.   The star-nose is a black, long-tailed mole. It is characterized by the twenty-two fleshy pink projections which fringe the tip of the snout (Figure 2.23). This singular structure immediately distinguishes *Condylura* from all other moles, and, for that matter, from any other mammal in the world. The forefeet are relatively weaker than those of other eastern moles, but are nevertheless broad and scaly, well adapted for a subterranean life. The fur is nearly black on the back, becoming paler on the sides and belly. The long tail is scaly and scantily haired, and often becomes enormously swollen in both sexes during the winter and early spring. Immature animals are much darker than the adults.

Average measurements of fifty-two adult males and females from central New York are: total length, 188 mm; tail, 66.5 mm; hind foot, 26 mm. The weights of fifty adults of both sexes from New York averaged 52 grams.

DISTRIBUTION.   The star-nosed mole is essentially a species of northeastern United States and southeastern Canada. It is found from southern Labrador to Manitoba south of James Bay, through part of Minnesota, northern Wisconsin, northeastern Indiana, Ohio, New York, and New England to Virginia, extending in the Appalachian Mountains to western North Carolina. Specimens have been secured in the Okefinokee Swamp of extreme southeastern Georgia. It is apparently scarce or absent in the coast region of the Carolinas (Figure 2.24).

HABITS.   The star-nosed mole is found in damp situations, be they meadows, fields, woods, or swamps. The animal delights in mucky pastures,

Figure 2.23. Nose of the star-nosed mole, *Condylura cristata.*

Figure 2.24. Distribution of the star-nosed mole, *Condylura cristata.*

or slow-flowing streams which have muddy bottoms. It is less often taken in the damp leaf mold of dense woods or in relatively dry fields which support a few damp spots from which its tunnels radiate.

Unlike its relatives, *Condylura* is as much at home in the water as on land. It is an adept swimmer, using its broad palms as efficient oars, while the tail is a useful scull. Frequently its burrows lead directly into a stream or pool. The tunnels are less regular than those of other moles, forming undulating subterranean highways which are frequented by meadow mice, shrews, and other small mammals. In mucky soil the mounds are similar to those made by crayfish.

The nest is usually composed of grasses, dead leaves, or such material as is available, and is constructed in some eminence which provides a measure of immunity from the periodic high water which often floods its chosen habitat. A compost heap is often chosen.

These moles are active throughout the year, pushing their tunnels through the snow and occasionally running on its surface. We have caught several in a minnow trap in eighteen inches of water, and in Indiana, two have been taken in muskrat traps a foot under water.

A single litter of three to six young is produced from mid-April to early June, after a gestation of about forty-five days. The young are blind and naked at birth but the nasal rays are visible. When three weeks old they are well furred and leave the nest to seek their own food and shelter.

Much of its food is secured on the stream bottom, where a variety of aquatic insects, crustaceans, and an occasional small fish are captured. The mole roots in the mud where aquatic worms, scuds, and a multitude of small forms live. The star-nosed mole is particularly attracted to stream borders in the winter. In the meadows and woods, earthworms provide their staple food (Hamilton 1931).

Predatory fish, hawks, owls, skunks, weasels, and cats are the chief enemies.

Condylura is seldom of any economic importance. Its very nature keeps it from the notice of the agriculturist and only rarely does it tunnel in lawns or on the golf green.

3

Chiroptera

(Bats)

Bats differ from all other mammals by virtue of their specialized forelimbs, which are adapted for flight. Other prominent characters include the backward direction of the knee, which is due to the rotation of the hind limb outward, and a cartilaginous process, the calcar, arising from the ankle joint and providing partial support for the interfemoral membrane. The ear is usually large with often a prominent tragus. Most bats have small eyes, and these probably serve only to distinguish light from darkness. The wonderful dexterity bats exhibit in avoiding obstacles and obtaining food is accomplished through echolocation. Flying bats emit supersonic notes and interpret the sound waves reflected back to them by objects, gaining information on the location, size, and movement of the object. Most bats of the north temperate regions exhibit delayed fertilization, with mating taking place in fall, but with fertilization occurring in late winter. There may be additional matings in the spring, or even in winter in some species. Originally, it was thought that this might have been an energy-saving device for the bats, but it may be more a matter of ensuring the fertilization of all bats.

Bats are world-wide in distribution, occurring in all the tropical and temperate parts of both hemispheres.

Key to Genera of Eastern Bats

(If the incisor is broken or dropped out, it will be represented by the tooth socket.)

A. Tail extending beyond interfemoral membrane.
    B. Upper incisors separated by conspicuous emargination of bony palate . . . .*Tadarida*
   BB. Upper incisors in contact with each other (recorded only from Florida, in eastern United States) . . . .*Eumops*

AA. Tail not extending beyond membrane.

    C. Ears very large, more than 25 mm in length, warty growths on muzzle . . . . *Plecotus*

    CC. Ears less than 25 mm in length.

        D. Interfemoral membrane wholly furred, or furred on basal half, above.

            E. Color blackish, hairs tipped with silver . . . . *Lasionycteris*

            EE. Color lighter, reddish, yellowish or silver . . . . *Lasiurus*

        DD. Interfemoral membrane not furred above.

            F. Only one upper incisor on each side of upper jaw . . *Nycticeius*

            FF. Two upper incisors on each side of upper jaw.

                G. Space behind upper canine tooth with one or two tiny teeth; five or six teeth in molariform tooth row (Figure 3.1).

                    H. Space behind canine with two tiny teeth; six teeth in molariform tooth row . . . . *Myotis*

                    HH. Space behind canine with one tiny tooth; five teeth in molariform tooth row . . . . *Pipistrellus*

                GG. No space behind canine; the four molariform teeth all of similar size . . . . *Eptesicus*

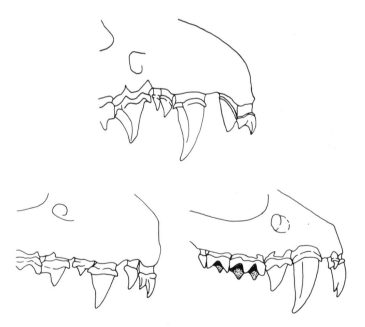

Figure 3.1. Toothrows of bats showing space behind canine with two small teeth in *Myotis* (top), space with one small tooth in *Pipistrellus* (bottom left), and no space, simply a large tooth behind canine in *Eptesicus* (bottom right).

Key to Bats of the Genus Myotis

Species of the genus *Myotis* are very difficult to identify, especially when preserved as study skins. The beginner may want to contact an authority for confirmation of identification. The calcar is the small bone projecting from the hind foot and partially supporting the interfemoral membrane. The tragus is the projection from the base of the ear.

A. Dorsal fur uniform in color throughout, rather than being dark at base; wing insertion is at the tarsus instead of at side of foot (normally seen only in fresh or alcoholic specimens .... *Myotis grisescens*
AA. Dorsal fur dark at base; wing insertion is at side of foot.
    B. Very small myotis (forearm 30–36 mm; hind foot generally less than 7 mm). Light colored but with black lips and mask. Keeled calcar . .*Myotis leibii*
    BB. Larger; no black mask. Calcar keeled or not.
        C. Tragus long (9–10 mm), thin and somewhat sickleshaped (Figure 3.6). Ears extending noticeably beyond nostrils when laid forward .... *Myotis keenii*
        CC. Tragus shorter and straight; ears extending about to nostrils when laid forward.
            D. Calcar keeled (seen only in fresh or alcoholic specimens). This bat is a relatively uniform pinkish brown coloration. In hibernation, bats hang in very tight, uniformly spaced clusters .... *Myotis sodalis*
            DD. Calcar not keeled; bats variable in color. May hang in tight clusters, but position of bats more variable.
                E. Dorsal fur dense, woolly; sagittal crest present in adults .... *Myotis austroriparius*
                EE. Dorsal fur normal, silky; no sagittal crest .... *Myotis lucifugus*

Little Brown Bat. *Myotis lucifugus* (Le Conte)

DESCRIPTION. The little brown bat has a furry face (Figure 3.2), only the nostrils and lips being naked. Its moderately long ears do not extend beyond the nostrils when laid forward. The wing membrane between the humerus and knee is sparsely furred, but the interfemoral membrane is not furred. The color above is a rich brown, at times almost bronze, but the hairs at their base are dark plumbeous or blackish. Ventrally the black fur is tipped with buff for the distal third. Young animals are much darker. Immature bats of all species can readily be distinguished from the adults by holding the wing membrane to the light. If the phalanges meet as knobs the animal is fully adult. The phalanges of young specimens gradually widen at their junction with one another.

Figure 3.2.   Head of little brown bat, *Myotis lucifugus.*

Average measurements of eighteen specimens of both sexes from Center County, Pennsylvania are: total length, 89.2 (79–93) mm; tail, 34.6 (31–40) mm; hind foot, 9.4 (8.5–10) mm; forearm, 38.1 (35–42) mm. The average weight is 4–5 grams during the spring and early summer. Just before hibernation individuals weigh 7.5–8.5 grams.

DISTRIBUTION.   The little brown bat occurs from Labrador to Alaska and in the east is found as far south as Georgia, Alabama, and Mississippi (Figure 3.3). In the northern part of its range it is the most abundant of all bats.

HABITS.   This species may be looked for almost everywhere. It is most commonly met with about towns and villages, from sea level to the forest mountain tops.

Shortly before dusk the little brown bat sallies from its daylight retreat behind a sheltering slab of loose bark, the dark recesses of a cave, or the attic of a house to flit on strong rounded wings over a nearby pond. After sweeping low over the water for a drink, it continues its flight, catching innumerable little insects with its tail or wing membranes and then deftly removing them with its strong teeth, or the bat may fly to a nearby tree where it can manage the larger prey with greater facility.

These little bats sometimes are seen flying in bright daylight, circling about little open glades in the forests where they undoubtedly find tiny diurnal creatures which provide them with food.

By late August and early September the bats have already put on a great amount of fat, amounting to about one-third of their body weight. At this time bats of many species may "swarm" or fly about the entrances of caves. A

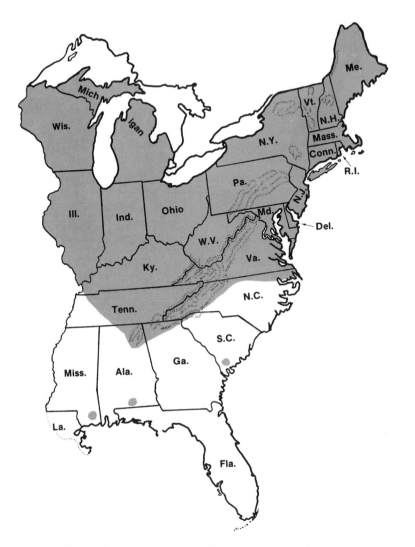

Figure 3.3. Distribution of the little brown bat, *Myotis lucifugus.*

mist net carefully placed over the entrance will often catch many in a night. The significance of this behavior is not known, as the bats seldom venture far into the cave; they simply circle about and into the entrance. Perhaps they are sampling air currents, thus gaining information about their possible use as hibernacula. As cold weather approaches, the bats begin to move haltingly toward the cave systems where they will hibernate during the winter. By late October little brown bats are entering their hibernacula. Here they hang by their strong feet, so closely packed at times that one can pick up a dozen in one's hands. At such times the bats feel cold to the touch and exhibit all the characteristics of true hibernation; but unlike a dormant ground squirrel, they soon awake and fly about if handled.

The females of most species of *Myotis* congregate, whereas the males are solitary; when males are found roosting with females, examination usually proves them to be immature animals. Several species, among them the little brown bat, are known to mate in the fall, the inactive sperm being stored in the uterus during the period of dormancy. When ovulation occurs in the spring, the now active sperm fertilizes the ovum. Gestation is about fifty to sixty days. There may be a second mating in the spring, before the bats have left their winter quarters, at which time fertilization occurs. Breeding colonies generally begin to form in April, often in attics of buildings.

A single young is produced from mid-June to mid-July. The newborn bat is blackish, with little fur, and remains hidden in some dark retreat while the parent forages during the evening. There is no evidence that the parent ever carries it on her hunting trips. It nurses for three or four weeks and first takes wing in the roost at about three weeks. The adult weight is attained by about four weeks, when they begin to fly outdoors and catch insects for themselves.

The little brown bat is wholly insectivorous. The stomachs of those which we have examined contained finely ground up remains of tiny beetles, bugs, and Diptera. Specimens shot about twenty minutes after they had taken flight had well-filled stomachs, a fact which suggests that these bats probably make several food-getting flights during the night.

Bats do not have many predatory enemies. Sometimes an individual cat, skunk, or raccoon will learn how to enter a cave and capture bats. We have seen many wings of bats inside the entrance to Wyandotte Cave which have fallen prey to a cat. Adverse climatic conditions probably exact a fearful toll. Successive nights of rain or high wind are sufficient to cause great mortality, for these little night flyers cannot long endure a fast. Several people have recorded this species caught on barb wire fences and burdocks.

These little bats are quite harmless and inasmuch as they destroy quantities of insects they should gain our gratitude. Unfortunately they often seek the dwellings of man, where their accumulated droppings give off a marked odor. Bats may be excluded from buildings by boarding up the place where they enter, or liberally sprinkling their retreat with several pounds of naphthalene flakes.

Southeastern Myotis. *Myotis austroriparius* (Rhoads)

DESCRIPTION. The southeastern myotis is similar in appearance to *Myotis lucifugus,* but the thick woolly and somewhat shorter fur and the absence of conspicuous burnished tips to the hairs distinguish it from the little brown bat. There is not a strong contrast in color between the tips and the bases of the hairs. The color above is dull yellowish or drabby brown, dull buffy below. Indiana individuals, once described as a separate subspecies, *M. a. mumfordi*, are white below, and are quite different from Illinois individuals.

The skull of this species is slenderer, with narrower interorbital constriction than in *M. lucifugus*. In addition, most skulls exhibit a low sagittal crest. This bat may be distinguished from other little brown bats of eastern United States by its large and strong hind foot without a keel on the calcar.

Average measurements of twenty adults of both sexes from Gainesville, Florida, are: total length, 91.5 (84.4–96) mm; tail, 41.3 (36.5–44) mm; hind foot, 10.7 (10.2–12.5) mm; forearm, 39.8 (35.5–41.7) mm. Males are slightly smaller than the females. Adult males weigh 5–7 grams; nursing females are heavier.

DISTRIBUTION.   This species has a disjunct distribution with large populations in the deep south, but isolated populations in southern Indiana and separate isolated populations in southern Illinois and Kentucky (Figure 3.4).

Figure 3.4. Distribution of the southeastern myotis, *Myotis austroriparius*.

HABITS.   The habits of this species are probably similar to those of the little brown bat. Careful studies of this species have been made at Gainesville, Florida, by H. B. Sherman (1930) and by Dale W. Rice (1937), whose observations are abstracted below.

Southeastern bats cluster together in caves or buildings or other protected sites such as hollow trees often in company with the free-tailed bats *Tadarida brasiliensis*. Some of the nursery caves in Florida contain thousands of individuals, but in other portions of the south caves are absent and buildings are used.

Nursery colonies begin forming in mid-March, building to huge numbers in Florida caves, 2,000 to 90,000 in some caves. During late April to mid-May, the two young (occasionally one) are born. The young, as they are born, are extruded into a pocket formed by the interfemoral and alar membranes, and crawl about in this for several hours after birth. They remain attached by the umbilical cord for several hours, and are only freed of this when the placenta is pulled clear and eaten by the parent.

At birth the young are naked except for a few vibrissae and a few hairs on the toes and about the knees. The ears and eyes are closed and except for the transparency of their skin, which gives them a pinkish hue, are black. During the day of their birth, the young remain attached to the mother, but later they hang in separate clusters from the adults except when nursing. They are capable of maintaining themselves when three weeks old. Most individuals leave the caves in winter and congregate in protected places, usually over water. Predation by rat and corn snakes appears to be an important mortality factor in the bat caves of Florida.

Gray Bat. *Myotis grisescens* A. H. Howell

DESCRIPTION.   The gray bat is similar in general appearance to *Myotis lucifugus* but slightly larger. This bat may be at once distinguished from all other members of the genus *Myotis* by the pattern of the dorsal fur, which is uniform throughout and not conspicuously darker at the base. Moreover the insertion of the wing membrane is at the tarsus instead of at the side of the foot, a character which is readily observed in fresh or alcoholic specimens, but is not evident in study skins. The dusky phase has uniform smoky fur above, below paler, the hairs "dark mouse gray" at their bases with dull whitish tips, those of the chin lighter, and those between the thighs whitish throughout. The russet phase is nearly uniform "cinnamon brown" above to the roots of the hairs, the lower surfaces with the whitish tips replaced by a pale buff, contrasting with the darker bases of the hairs (Miller and Allen 1928). The skull is larger than in *Myotis lucifugus* and has evident sagittal and lamboidal crests.

Average measurements of twenty-five adults from various parts of the range are: total length, 88.6 (80.2–96) mm; tail, 38 (32.8–44.2) mm; hind foot, 9.9 (8.4–11.2) mm; forearm, 43.1 (40.6–45.8) mm. Mohr found these bats to weigh 7.64 to 9.1 grams during June; January specimens weighed 6.47 to 7.20 grams.

DISTRIBUTION.   This species is found throughout the limestone region of the southern middle west to the southeastern states; i.e., from southern Illinois (one summer colony present) to Arkansas, Missouri, Alabama, and Western Florida (Figure 3.5). This species still occurs in very large numbers, but it is in need of protection and has recently been placed on the federal endangered species list because of its habits of congregating in large numbers in so few caves, especially in winter. Disturbance of a relatively few caves could wipe out vast portions of the population of this species.

HABITS.   Both *Plecotus townsendii* and the gray bat may be considered true cave bats, for they are seldom found far removed from their winter quarters or from the caves in which the young are born. Gray bats are highly migratory, spending the summers in fairly large colonies in several states but

Figure 3.5. Distribution of the gray bat, *Myotis grisescens.*

especially in Missouri, Tennessee, and Kentucky. They move between summer and winter quarters in large flocks. In 1970 one summer colony in the Shawnee National Forest of southern Illinois had about 1000 to 2000 individuals. Banding studies by John Hall and Nixon Wilson (1966) indicate that all or most of the bats from summer colonies in Tennessee, Kentucky, and Illinois hibernate in huge colonies in one large cave complex, Coach-James Cave in Kentucky. The hibernating cave complex is not used by gray bat colonies in summer.

Even though the gray bat frequents caves throughout the year, there is a remarkable segregation of the sexes during the season of reproduction, for the breeding females seem to congregate apart from the males. Breeding caves are usually large, and contain much water. Breeding colonies begin to form in late March and early April; colonies break up in late July and August, after the young are weaned. They reach the hibernating caves in October. The species probably mates in the fall, although Mary J. Guthrie cites evidence which suggests that copulation normally occurs in the spring as the bats are aroused from hibernation, and ovulation normally follows this spring mating.

The young are born from early June to early July, depending on the latitude. At birth the baby weighs about a third as much as the parent. The mother carries the naked helpless young for several days after birth, but later leaves it hanging from the ceiling or wall of the cave. The young are able to provide for themselves when three weeks old.

Many young drop from the walls, or lose their grip on the flying mother to fall on the cave floor and perish. Scores of dead young may be seen beneath extensive bat roosts during early summer.

A. H. Howell (1921), writing of *Myotis grisescens* in an Alabama cave, speaks of their hanging from the ceiling in one compact mass covering three or four square feet and several bats deep. During mid-June the bats were observed coming out of this cave about 7:00 P.M.; they swarmed out in large numbers, feeding in the mouth of the cave and among the trees on the river bank. Quoting Ernest L. Holt, a visitor to the Rogerville Cave in northern Alabama, Howell writes:

"The bats were not hanging in clusters, but thousands of them lined the ceiling in a solid sheet, hanging separately head downward. A couple of shots were sufficient to cause pandemonium, and immediately I found myself in almost total darkness (my assistant having retreated around a corner with the lantern), surrounded by a swirling mass of squeaking bats. They were everywhere, the flying thousands filling the air, and in their panic rushing against me and sticking all over my head and body; I had to keep kicking to prevent them crawling up my breeches legs. On my way out the scattered bats seemed to fill the whole cave. Their droppings covered the floor in places to the depth of several feet."

Clark, LaVal, and Swineford (1978) found brains of juvenile gray bats dead beneath maternity roosts in two Missouri caves contained lethal concentrations of dieldrin. One colony appeared to be abnormally small, and more dead bats were found a year after the juvenile bats had been collected. This is the first report to link the field mortality of bats directly to insecticide residues acquired through the food chain.

Keen's Bat. *Myotis keenii* (Merriam)

DESCRIPTION.   This species is similar in appearance to the little brown bat, but is distinguished by its long ears (about 14 to 18 mm from the notch), which, when laid forward, extend about 4 mm beyond the nostrils. The tragus in this species is long (about 8 to 10 mm from the notch), narrow and somewhat curved (Figure 3.6). In color it is similar to *Myotis lucifugus,* but the brown hair tips of the dorsum are neither so long nor so glossy. The venter of this species is usually yellowish. The two species must be laid side by side for such color differences to appear. The skull is narrower in proportion to its length than that of *Myotis lucifugus.*

Average measurements of twenty adults from various parts of the range are: total length, 84.1 (79.2–87.8) mm; tail, 41.7 (36.4–43) mm; hind foot, 8.4 (7.2–9.4) mm; forearm, 36.7 (34.6–38.8) mm; weight, 4.6–7 grams.

Figure 3.6. Tragus of Keen's bat, *Myotis keenii.*

Figure 3.7. Distribution of Keen's bat, *Myotis keenii.*

DISTRIBUTION. In eastern United States this species occurs from Maine to Wisconsin, south to northern South Carolina and Georgia, thence westward through most of Tennessee (Figure 3.7).

HABITS. Keen's bat is far more abundant than specimens in collections would indicate. During the day it hangs in small colonies of several to a dozen or more, selecting a convenient window blind, loose bark, or a cave. During mid-August in a New York cave a single Keen's bat was found hanging in the midst of twenty little brown bats, its long ears alone indicating that it was a different species. Just at dusk on an August night, R. E. Mumford and Whitaker set up a mist net just inside the entrance of Ray's Cave in Indiana. It was so full of Keen's bats that we had to remove it within ten minutes. Hamilton once broke a dead yellow birch stump in a Maine bog and found a Keen's bat tucked away for the day. Another was found hanging in full daylight on the front porch of a cabin on a lake in Clay County, Indiana. Its habits do not differ materially from those of the little brown bat, although individuals of this species probably lead a more solitary life for they are more often found alone, and in caves usually hibernate alone or in very small groups.

This bat is quite common in New York. One summer in a little cottage on the shores of a lonely lake in eastern New York, several bats of this species had taken up their abode behind the shutters, and could readily be studied. They usually left quite late in the evening, and seldom returned until dawn. It is quite possible they had other quarters to which they repaired when they became tired of hunting. As they flew into the daytime retreat, they would make sundry high-pitched squeaks, until the rising sun put an end to their movements. They did not appear to sleep soundly, for however slowly and carefully the shutters were pulled back, they would immediately awaken, and would back over the shingled siding and fly off if further disturbed. Newspapers were placed below the blinds to catch the shiny black scats, and when these were examined they revealed finely ground up remains of caddis flies, mayflies, and Diptera.

The Keen's bat has a single young which is born somewhat later than those of other bats. We have taken females with a large embryo in New York State on June 21, June 29, and July 9. It thus appears probable that the season of parturition comes in July, but very few breeding data are available for this species.

Indiana Bat. *Myotis sodalis* Miller and Allen

DESCRIPTION. This little bat bears a strong general resemblance to *Myotis lucifugus,* which probably explains why it was not described as a distinct species until 1928.

The texture of the fur is extremely fine and fluffy; the hairs have a tendency, due perhaps to a slight crinkling, to stand out from each other a little, as in *Pipistrellus subflavus.* In comparison with the bronzy burnished-tipped fur of *Myotis lucifugus* the pelage of this species is dull grayish chestnut. The color is distinctive. On the upper surface the basal two-thirds of the hair is fuscous-black; then comes a narrow grayish band succeeded by a cinnamon-brown tip, so that there is a distinctly tricolor effect, while the grayish band showing through the cinnamon-brown tips gives a peculiar hoary appearance at a short distance. Below, the fur is slaty basally, the hairs with grayish-white tips, washed more or less heavily with cinnamon brown particularly at the flanks, instead of slightly yellowish as in *M. lucifugus*. The general effect is a pinkish white below and a dull chestnut gray above (Miller and Allen 1928). The skull usually has a slight sagittal ridge but the teeth are indistinguishable from those of *M. lucifugus* (Figure 3.8). An excellent character is the well-developed keel on the calcar, but unfortunately this usually does not show on study skins.

Average measurements of thirty-six specimens from various parts of its range are: total length, 81.7 (70.8–90.6) mm; tail, 36.4 (27–43.8) mm; hind foot, 7.9 (7.2–8.6) mm; forearm, 38.8 (36–40.4) mm; weight 5–8 grams.

DISTRIBUTION.   The Indiana bat occurs throughout eastern United States from the Central Mississippi Valley and northern Alabama to New England (Figure 3.9).

HABITS.   Most specimens of this bat have been collected in caves during the winter months, where the species clusters in large, rather characteristic masses, tightly packed, and often all lined up one deep and in similar direction, so that the pink faces of all show. These hibernating animals come

Figure 3.8.   Head of Indiana bat, *Myotis sodalis.*

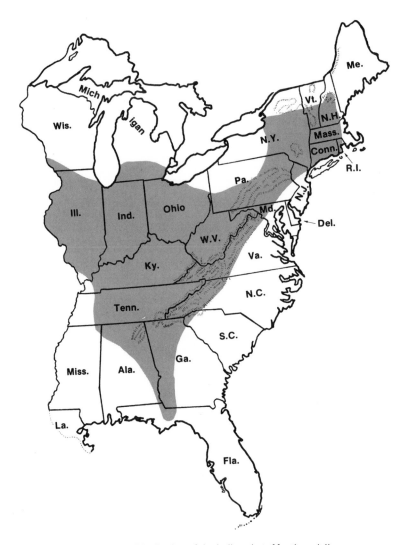

Figure 3.9. Distribution of the Indiana bat, *Myotis sodalis.*

together from a considerable area, and it is precisely because they congregate in so few caves in winter that this species is on the federal endangered list, even though still quite abundant. The loss of a few hibernating populations would greatly influence the total population. Collectors should give careful attention to all little brown bats taken during the summer months, for the species has been recorded from only a relatively few localities in its wide range. Few individuals have been found clustering together in barns or behind shutters as *Myotis lucifugus* is accustomed to do.

Myotis sodalis appears to favor caves in which there are considerable bodies of water. In such caverns it often forms great masses on the walls and ceilings. The colonies are composed largely of males. Indeed males appear to preponderate in these winter quarters for all bats. In a collection of 781 specimens of *Myotis sodalis* in Penn's cave reported by Charles E. Mohr (1933), a count showed males to constitute 79 percent of the total.

This bat and the gray bat are host to a peculiar spider mite, *Spinturnix globosus,* normally found in the anus of the host.

That this species spreads out over a considerable area from its winter hibernating quarters is evidenced by the number of individuals taken in summer in areas where no caves occur. The bats leave the hibernacula in March. We have shot a number of Indiana bats in summer at Terre Haute, Indiana, and numbers from throughout the state are turned in to the Indiana Health Department each year for testing for rabies. Like many other bats, this species exhibits delayed fertilization. Breeding occurs in the fall after the bats have entered the hibernacula. Hundreds can be seen copulating at night at Bat Cave, Kentucky, in early October. Until recently, very little was known of the breeding habits of this species, as only three pregnant individuals had been taken. James B. Cope and associates at Earlham College now have under study three breeding populations of this species. These bats form breeding congregations under loose bark of trees along wooded Indiana streams.

Small-Footed Bat. *Myotis leibii* (Audubon and Bachman)

DESCRIPTION.   The small-footed bat is similar in appearance to the olive phase of *Myotis lucifugus*. It can readily be distinguished from the latter by its smaller size, black ears and facial mask, golden tinted fur, keeled calcar, and shorter forearm. The face from the nose to the base of the ears and including the lower lip is black, giving a masked appearance, which is heightened by the dull black ears, tragus, nose, and chin. The wing and membranes are blackish brown (Figure 3.10). The skull is nearly as long as that of *M. lucifugus*, but is so flattened that it has noticeably less depth, the brain case being distinctly narrower. Average measurements of six specimens from New York, Vermont, and Maryland are: total length, 77.5 (73–82) mm; tail, 32.6 (29.8–35.2) mm; hind foot, 6.8 (6.6–7) mm; forearm, 31.8 (30.8–34) mm.

DISTRIBUTION.   The small-footed bat occurs from western New England to Kentucky and West Virginia. It is one of the rarest of bats in collections,

Figure 3.10.   Head of the small-footed bat, *Myotis leibii.*

but will undoubtedly become better known when collectors become acquainted with its characteristics (Figure 3.11).

HABITS.   The only detailed studies of Leib's bat have been made by Charles E. Mohr (1936), who has collected and marked a good series of this species. All were found in caves in a very limited area during the winter. Mohr believes that this species is restricted to a very definite ecologic situation. The caves of central Pennsylvania where the small-footed bat occurs are in wild mountainous country rising to 2,000-foot altitudes. The caves themselves are in unusual situations, being located in hemlock forests. Wayne H. Davis (1955) has found several Leib's bats on the ground under rocks in caves. He reports that in one West Virginia cave, they hibernate in crevices in the cracked clay floor.

Mohr writes:

"Another observation on the habits of this bat that seems significant is that it is not only comparatively abundant but also very active in late winter and deserts the caves long before the other hibernating bats. I believe that this indicates a migratory movement in late winter. With scarcely an exception the least bats taken responded quickly when touched, and a number which were seen escaped before they could be caught. Indeed, during the past few months we saw at least a dozen of these bats which could not be reached with the fishing pole which we used, or which flew as we attempted to catch them. The attitude of the bat as it hangs on the walls or ceilings of the caves is so characteristic that it offers a very definite means of identification at a distance.

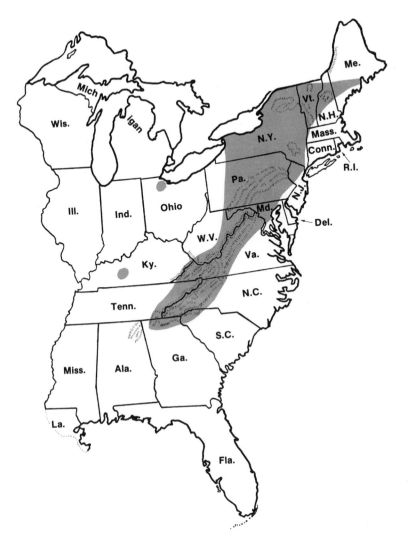

Figure 3.11. Distribution of the small-footed bat, *Myotis leibii.*

The arms instead of hanging practically parallel are extended about 30° from the vertical, and almost every least bat was found in this position, other bats seldom. The last least bat was taken this year (1933) on April 2. No caves were visited the following week but visits to Dulany Cave on April 15, Cornwall Cave on April 16 and Aitkin, Little Aitkin, and Stover Caves on April 17, failed to show a single bat of this species, although other hibernating bats were still inactive.''

Mohr believes that *Myotis leibii* hibernates in caves of the Allegheny Mountains from Vermont to West Virginia, migrating southward during February and March or possibly earlier and stopping over for short periods in remote caves in heavily forested sections of the mountains. Specimens have been shot about large caves in Kentucky during August. Little is known of breeding, but presumably one young is produced. Nursery colonies of twelve to twenty bats have been found in buildings in the West.

Silver-Haired Bat. *Lasionycteris noctivagans* (LeConte)

DESCRIPTION.   The silver-haired bat can at once be distinguished from all other eastern bats by its color. The pelage in both sexes is alike, being dark blackish brown; the ends of the hairs are tipped with silver. This frosting is most pronounced along the middle of the back and is absent, or nearly so, from face, crown, and throat. The fur extends on the upper surface of the interfemoral membrane for half its length. The short rounded ear, nearly as broad as long, has a broad, bluntly rounded tragus. The flattened skull has a broad rostrum, the dorsal profile being practically straight (Figure 3.12).

Average measurements of ten specimens from New York are: total length, 105.8 mm; tail, 42.4 mm; hind foot, 7.9 mm; forearm, 41.1 mm; expanse, 290–310 mm; weight, 7–9 grams.

Figure 3.12. Head of the silver-haired bat, *Lasionycteris noctivagans.*

Figure 3.13. Distribution of the silver-haired bat, *Lasionycteris noctiva-gans.*

DISTRIBUTION.   *Lasionycteris* is peculiar to North America, the single species occurring from the Atlantic to the Pacific. In eastern United States this bat occurs from Maine and Wisconsin to northern Georgia, but probably does not commonly occur south of the Smoky Mountains during summer. Possibly it winters in the Gulf States (Figure 3.13).

HABITS.   This is a bat of the watercourses, flying about the trees and tall shrubs which border streams and lakes. It leaves its diurnal retreat early in the evening, often being abroad before the sun has fully set. It is easily recognized in flight by its slow and erratic course. The silver-haired bat is a common species of northern regions; indeed, C. Hart Merriam (1884) states that it is the most abundant bat of the Adirondacks.

It is usually solitary, but one huge colony was found in a hollow tree. More often a single individual may be found behind a piece of loose bark or hanging from a twig in some dense hemlock stand. Like many other species of bats, there is a marked segregation of sexes. Out of eighty-five adult specimens killed by Merriam in Lewis County, New York, during the summer of 1883, there was but a single male. This sexual discrepancy does not exist at birth. In all probability the males live a solitary life about the little forest ponds and lakes where bat collectors have not yet ventured. Or they may pass the summer months considerably farther north than the breeding females.

This species is a fearless little creature. Success has been attained in collecting individuals with a handful of alder switches as they repeatedly flew past on their hunting paths bordering a little northern bog pond.

When this bat is shot and falls into the water, it swims rapidly and strongly to shore if its injuries permit.

In northern latitudes the young, which number usually two, are born in late June or early July. At birth they are black wrinkled objects, and they remain clinging to their quarters while the parent forays for food. The bats grow very rapidly, and when three weeks old are strong enough to follow their parents on the nightly quest for food. They can easily be distinguished from the adults at this season by their relatively weak and hesitant flight; and, if one is captured, the pronounced silvery-tipped, almost black hairs, serve to distinguish it from the duller adults.

As cold weather approaches, *Lasionycteris* leaves its northern home and journeys in a leisurely fashion southward. At such times the bats may make long journeys over the ocean, for they have been seen far from land flying beneath heavy storm clouds several hundred feet above the choppy sea. When forced to rest, they congregate in some numbers on boats, hiding in the furled sails, hulls, and cabins of yachts. Indeed, they may pass the winter in such places. These bats have been found hibernating in skyscrapers, churches, wharf-houses, and the hulls of ships in New York City, and in the silica mines of southwestern Illinois during the months between December and March, and a specimen has been found hibernating beneath the loose bark of a tree in British Columbia. Individuals have been found in hollow trees (perhaps the most common hibernating site) in North Carolina during the winter months, and there are records for Bermuda.

Some years this bat is rather common, while at other times it may be quite scarce. It is a species that will repay close study.

Eastern Pipistrelle. *Pipistrellus subflavus* (F. Cuvier)

DESCRIPTION.   This little bat is readily distinguished from all other eastern bats by its small size, tri-colored fur, and dentition. The dorsal hairs are
deep plumbeous at their base, then yellowish-brown almost to the tips, which
are dark brown. The fur is light uniform yellowish-brown below. Considerable variation in color occurs, but the general overlying tone of the dorsum is
gray rather than brown. The dorsal base of the interfemoral membrane is
sparsely furred. The extremely thin and delicate wing membranes are attached
to the base of the toes. The thumb is large for such a small bat, being about
one-fifth the forearm. The ears are distinctly longer than broad and taper to a
narrow round tip (Figure 3.14). It has a total of thirty-four teeth, including
five molariforms on each side of the upper jaw, with the first reduced, creating a space between the canine and second molariform. Bats of the genus
Myotis have two teeth in this space, while only *Pipistrellus, Plecotus,* and
Lasionycteris have one tooth in this space (Figure 3.1).

Average measurements of seven specimens from Georgia are: total length,
85.1 (81–89) mm; tail, 40 (36–45) mm; hind foot, 8.7 (8–10) mm; forearm,
33 mm; expanse, 245 mm; weight, 3.5–6 grams.

DISTRIBUTION.   The pipistrelle occurs in the eastern United States from
northern New England south to central Florida and west to Wisconsin (Figure

Figure 3.14. Head of eastern pipistrelle, *Pipistrellus subflavus.*

3.15). Northern pipistrelles, originally described as *obscurus,* are darker, duller, and more yellow. Specimens from central New York are decidedly more yellow or light brownish than those from Georgia. Some specimens are very dark brown.

HABITS.   This small bat is a weak, erratic flyer, usually following an undulating course. It has been mistaken for a large moth, and we have,

Figure 3.15. Distribution of the eastern pipistrelle, *Pipistrellus subflavus.*

indeed, nearly succeeded in capturing a flying individual in an insect net. These bats are early flyers, coming forth from a building, a cleft in a cliff or other retreat to course over the water and flit about shaded groves in the early evening.

The bats are sociable little creatures and often remain together for several years. A. A. Allen (1921) banded a cluster of four adult females at his Ithaca, New York, residence in 1916 and recaptured them in the same place three years later.

The pipistrelle remains active late into October, but as food becomes scarce the little bats, now very fat, retreat to some cave to pass the winter. This species will hibernate in very tiny caves or mines, often being found when no other bat species is present. It was found in all thirteen caves having bats that we examined in winter 1974–75 in a study on the Shawnee National Forest in southern Illinois, and in eight of the eleven silica mines inhabited by bats. Unlike other cave bats, which often move about in their winter quarters, the pipistrelle remains really dormant, seldom moving from the site it has first chosen. If the cave be moist, water collects on these dormant sleepers until they become covered with little droplets. When the beam from a flashlight catches these little figures they fairly glisten in the light. At such times they appear to be white. The first warm days of March often stir them to activity in the north and they leave the caves during the day, to fly in the bright sun until the chill of nightfall sends them back. In the south some individuals are thought to remain active throughout the winter. Little is known of the summer roosting sites, but several nursery colonies are known from buildings, and one was found in a barn at Terre Haute. Trees are frequently selected as summer daytime roosting spots.

From late June to mid-July in northern latitudes, the females produce two young. These are cared for solicitously, and are carried by the parent on her evening flight during the first few days of their life. In the southern states the young are produced in late May through mid-June. After a few days the young are left alone while the mother seeks food, and when they are less than three weeks old they too take to the air. Young at birth are quite large for such a small bat, a litter of two weighing 1.89 grams.

As with many other bats, the segregation of sexes is pronounced during the season of parturition. During August, bats of both sexes may enter the mouths of caves long after nightfall, often described as swarming. Large numbers of bats can often be caught in mist nets at this time. Poole (1938) suggests that these nocturnal flights may have a sexual object. At any rate, it is the only time at which both sexes are to be found together in any numbers.

From a microscopic study of their scats, Hamilton concluded that their chief food consists largely of tiny flies, beetles, and hymenopterous insects. Specimens which he shot in eastern Kentucky fifteen to twenty minutes after they first appeared in the evening had their stomachs greatly distended with small dipterous and coleopterous remains. This suggests that these bats do not remain on the wing throughout the night but make an early evening flight, then perhaps another toward midnight or early morning. Twenty-three individuals from Indiana had cicadellids (21.7 percent volume), carabids (18.1 percent), unidentified flies (10.7 percent), unidentified beetles (7.8 percent), moths (7.3 percent), and delphacids (7.2 percent) as their primary foods.

Big Brown Bat. *Eptesicus fuscus* (Beauvois)

DESCRIPTION.   This bat is at once distinguished from all other eastern bats by its uniform sepia-brown long fur and large size. The fur of the ventral surface is much paler. The hairs are much darker at their base. The short black ears are furred at the base. Fur is lacking on the wing and the interfemoral membranes (Figure 3.16).

Average measurements of twenty specimens from New York and Georgia are: total length, 114.3 (106–127) mm; tail, 46 (42–52) mm; hind foot, 10.7 (10–11.5) mm; forearm, 46.8 (44.7–48) mm; expanse, 320 mm; weight, 13–18 grams. Large females, with embryos removed, may weigh to 30 grams.

DISTRIBUTION.   The big brown bat occurs throughout most of the United States and adjoining Canadian provinces. In the east it ranges from Maine and Wisconsin southward to central Florida and southern Louisiana (Figure 3.17).

Figure 3.16.   Head of big brown bat, *Eptesicus fuscus.*

Figure 3.17. Distribution of the big brown bat, *Eptesicus fuscus.*

HABITS. The big brown bat, easily recognized by its general brown color, large size, and strong erratic flight, usually takes to the air somewhat later than the smaller species. It passes the day either singly or in small clusters in the recesses of some building, hollow tree, or cave. The big brown bat often resorts to the habitations of man. If a bat enters and flies about the house, the chances are rather good that it is this species. It is not dismayed by the bustle and noise of the great cities, for frequently individuals have been seen flying above the crowded streets of New York City.

Eptesicus is one of the last bats to disappear in the fall; during mild spells of winter it may be seen flying in the sun at midday even in the northernmost parts of its range. A number of big brown bats appear in winter on the Indiana State University campus in Terre Haute, Indiana, and many are found in buildings throughout Indiana, and are submitted to the Indiana department of health for rabies examination. However, the species does not congregate in winter. Rather each individual finds its own little nook or cranny and jams itself in. The partition, eaves, or attic of a house is often selected as the hibernation site of this species. Even exposed window sashes have been chosen for a place in which to pass the winter, and Whitaker found one hibernating on the floor of an old building under a pile of lumber. Many big brown bats hibernate in caves in the company of *Myotis* and *Pipistrellus*. However, *Eptesicus* hangs singly or usually in clusters of not more than two to four, and often very close to the cave entrance.

As in many other hibernating bats, this species exhibits delayed fertilization. Copulation is known to occur in spring or fall, but Russell E. Mumford has observed several matings in caves during the winter period. There is a good deal of activity by some individuals of this species during the winter. Its function may be to ensure fertilization.

The season of partus occurs in mid-June in the north, several weeks earlier in the south. Two young are the usual number in the east, although Francis Harper has collected a Georgia specimen containing four embryos. Vernon Bailey obtained bats in Washington, D.C., on July 31 and August 7, each of which contained two minute embryos the size of No. 8 shot. He concluded that they would have been born in May or June of the next year. This is a most unusual situation, for fertilization of the egg presumably does not occur until early spring in most species of bats. At birth they weigh 2.8 to 3 grams. The young grow rapidly and are weaned when three weeks old. When two months old they are as large as the parents.

An examination by Hamilton (1933) of 2,200 fecal pellets of *Eptesicus* indicate the following food: Coleoptera, 36.1 percent; Hymenoptera, 26.3 percent; Diptera, 13.2 percent; Plecoptera, 6.5 percent; Ephemeridae, 4.6 percent; Hemiptera, 3.4 percent; Trichoptera, 3.2 percent; Neuroptera, 3.2 percent; Mecoptera, 2.7 percent; Orthoptera, 0.6 percent. No lepidopterous remains were found. Of the beetles, the family Scarabaeidae (leaf chafers and May beetles) occurred most frequently; next in order occurred the family Elateridae, in which the wire worms are placed. The Diptera were represented chiefly by muscids, a group embracing the house fly and its allies. Whitaker's examination of 184 stomachs of this species from Indiana gave similar results. The most important foods were Carabidae, Scarabaeidae, Chrysomelidae, Pentatomidae, and Formicidae, forming percentage volumes as follows: 14.6, 12.4, 11.5, 9.5, and 8.5. Many of the prey items eaten were fairly large, being at least a centimeter long. Beetles amounted to 49.6 percent of the total volume in the stomachs. Only 4.5 percent was of moths.

Bats have few enemies, but it is known that barn owls capture a few. A pilot black snake has been responsible for the decimation of a colony of these bats. The snake took up its residence among the eaves of a two-story dwelling and fed on such bats as came within reach.

Key to Bats of the Genus Lasiurus

A. Interfemoral membrane furred on basal half only. Color yellow . . . .
. . . .*Lasiurus intermedius*
AA. Interfemoral membrane wholly furred above.
   B. Forearm more than 45 mm. Dorsal hairs mostly tipped with white . . . . . .
. . . .*Lasiurus cinereus*
   BB. Forearm less than 45 mm. Color mahogany brown, red, orange or yellowish.
      C. Color red, orange or yellowish . . . . *Lasiurus borealis*
      CC. Mahogany brown . . . .*Lasiurus seminolus*

Red Bat. *Lasiurus borealis* (Müller)

DESCRIPTION.   Its conspicuous bright reddish or rusty color at once distinguishes this bat from all other species. The low broad rounded ears with triangular tragus are naked on the inside but the upper sides are densely furred for their basal two-thirds. The interfemoral membrane is thickly furred on its upper surface but only thinly furred on its proximal ventral surface. The fur also extends along the under surface of the wing along the humerus to the wrist. The hair tips of the dorsum and breast are frosted with white. This species is one of the very few bats in which the sexes are contrastingly colored. Females have a dull buffy chestnut coat much frosted with white, whereas the males are much brighter, almost orange red. Each sex has a buffy white patch on the front of each shoulder (Figure 3.18).

Average measurements of ten adults from New York, Pennsylvania, and Georgia are: total length, 112.3 (95–126) mm; tail, 49 (45–62) mm; hind foot, 9.2 (8.5–10) mm; forearm, 40 (37.5–42) mm; expanse, 330 mm; weight, 9.5–15 grams.

DISTRIBUTION.   The common red bat occurs over all of eastern United States, ranging from Canada to northern Florida and westward to Colorado (Figure 3.19). It is very abundant in the midwest and apparently is the most abundant bat in summer in Indiana.

HABITS.   This common and handsome species is the most beautiful of all American bats. Its exquisitely soft, fluffy fur is in marked contrast to the oily finish of the little brown bats (*Myotis*) which inhabit caves. The red bat chooses a branch of some shady tree in which to pass away the daylight hours.

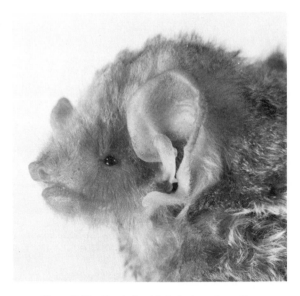

Figure 3.18. Head of red bat, *Lasiurus borealis.*

It may rest within a few feet of the ground, usually partly concealed in a mass
of leaves. We have found them attached to corn stalks in full sunlight.

Long before darkness has fallen, the red bat is abroad on strong narrow
wings, pursuing its swift erratic course. On still summer evenings these bats
may be seen at great heights, but they soon spiral down, at times flying but a
few feet above the ground as they course for their favorite prey.

On late summer evenings these handsome bats fly about the street lights of
the city, attracted no doubt by the myriads of insects which are drawn to the
lights. At times they will alight on the supporting poles to pick up moths
which have come to rest.

Examination of the stomachs of 128 bats of this species from Indiana
revealed that moths were a very important food in the diet, although a great
number of other foods were eaten. Moths formed 26.2 percent of the food in
the sample, by volume, whereas June bugs, planthoppers (Delphacidae), ants,
leafhoppers, unidentified beetles, and ground beetles were the other important
foods. Beetles, collectively, formed 28.1 percent of the volume in the sample.
Red bats often fail to capture their prey on the first attempt. One was seen to
make ten unsuccessful attempts to capture a glittering miller in the full glare of
an electric street light. In August and September red bats are often caught at
mist nets set at cave entrances, along with several other species of bats.

The red bat mates during August, somewhat earlier than most bats. Copulation occurs while the bats are in flight. We have collected females the first
week of August which contained quantities of sperm in the uterus. The three

Figure 3.19. Distribution of the red bat, *Lasiurus borealis.*

or four young are born about the middle of June, and there is one case of five in a litter. These are high numbers for bats, and the females possess four teats with which to nurse this number.

It has been said that when the young are small, they accompany the parent on her flights, grasping the teat or loose skin of the breast with their tiny recurved milk teeth. However, we have shot and netted a number of female bats in Indiana without finding any young attached. We have found a number of females on the ground with young attached. The recurved milk teeth may function to help the young remain attached to the female when she is blown or frightened from her arboreal perch. This would enable her and the young to glide to the ground, and when danger or the wind is past, then to climb back up a tree. Greater predation than on many other species of bats would be indicated by the higher numbers of young. Soon the combined weight of the young exceeds that of the mother, but they continue to nurse for some time. The young at this time hang by their feet but retain contact with the mother, grasping her with folded wings while her expanded membranes give them a measure of shelter. A month after birth they are large enough to forego parental care and commence their own solitary life.

From September until late November, most red bats presumably move to more southern latitudes to pass the winter. It is our best known migratory species. Individuals have been observed several hundred miles at sea and on Bermuda. However, there are almost no data on movements of individual bats, and few data from specimens in collections. Detailed analysis on dates and sex of red bats in collections should help us to arrive at hypotheses concerning migratory movements. Just where the winter is spent is not known, but inasmuch as the red bat becomes more abundant in the southern states from December to March, it is presumed that these are their winter quarters. However, red bats commonly winter at least as far north as Kentucky, Indiana and Illinois. Wayne H. Davis and William Z. Lidicker (1956) observed red bats emerge from woods in November and Davis reports counting twenty-seven over roads in Mammoth Cave National Park, Kentucky, on warm afternoons in January 1957. Russell E. Mumford has seen red bats flying on warm afternoons in Southern Indiana. Presumably these individuals enter torpor at these localities; they arouse from torpor at 59°F or at much higher temperatures than hibernating cave bats. This would protect them from waking too frequently, thus wasting energy during the winter. Some temperature regulation does occur which keeps them from freezing.

Seminole Bat. *Lasiurus seminolus* (Rhoads)

DESCRIPTION.   The Seminole bat is similar to *Lasiurus b. borealis* but much darker, the rufous shades of the northern species replaced by rich mahogany brown, slightly frosted with grayish white. Measurements of six adult specimens average: total length, 111 (108–114) mm; tail, 48 (44–52) mm; hind foot, 8.5 (8–9) mm.

DISTRIBUTION.   The Seminole bat ranges from North Carolina through Georgia, Florida, Alabama, and eastern Louisiana (Figure 3.20). A specimen has been taken in Berks County, Pennsylvania, and another at Ithaca, New York.

Figure 3.20. Distribution of the seminole bat, *Lasiurus seminolus.*

HABITS.   The best account of this species has been given by Francis Harper (1927) who studied it in the Okefinokee Swamp of southeastern Georgia. He writes:

"In winter this appears to be the commonest bat of the swamp, but in summer it is outnumbered by Rafinesque's Bat. . . . The feeding grounds of this species appear to be very largely over watercourses, pine barrens, and cleared land, and to a less extent over prairies and hammocks. . . . In summer they likewise appear during the last half hour before deep dusk. It seems to fly more directly than most of the other local species, and sometimes, at least, it travels comparatively slowly.''

In southern Georgia and Florida parturition occurs toward the last of May. Three or four young are the customary number (average of 3.3), and they are supposed to be capable of flight at an age of probably no more than three weeks.

These commonly roost in long bunches of Spanish moss through most of the year. The clumps range from about 3.5 to 15 feet above the ground. The area below the clump is always free of limbs, this allowing the bat to drop free into flight.

Apparently the Seminole bat is active throughout the winter, but it is more in evidence on warm evenings. It may occasionally descend to the ground to feed, for H. B. Sherman (1935) shot a specimen at Gainesville, Florida, which had in its jaws a flightless cricket.

There are records of this bat in southeastern Pennsylvania and in New York, more than three hundred miles north of its normal range.

Hoary Bat. *Lasiurus cinereus* (Beauvois)

DESCRIPTION. This grand bat is the largest of eastern species. Its yellowish-brown to dark mahogany-brown hair is frosted with silver over the entire body, giving a pronounced hoary appearance to the animal. The short rounded ears have black naked rims. The top of the feet and the entire interfemoral membrane are well furred. The throat and wing linings are buffy. The texture of the fur is unusually full and soft. Average measurements of six adults from various parts of its range are: total length, 135 mm; tail, 59 mm; hind foot, 13 mm; forearm, 52 mm; wingspread, 400 mm. Adults weigh from 20 to 35 grams, the females usually being the heavier (Figure 3.21).

Figure 3.21. Head of hoary bat, *Lasiurus cinereus.*

Figure 3.22. Distribution of the hoary bat, *Lasiurus cinereus.*

DISTRIBUTION. The hoary bat occurs over all of temperate North America, extending north of the Arctic Circle in summer and passing the winter south of the latitude of New York. Specimens have been collected in Mexico. In the Gulf States it is probably a winter resident only (Figure 3.22).

HABITS. As darkness falls on some lonely northern lake, the bat hunter strains his eyes to pierce the deepening gloom. A moment before he has seen a great bat fly swifly over the water on strong narrow wings, its erratic flight preventing a possible shot. It is the great hoary bat, a prize that falls to few collectors. Most specimens are taken by hand in an unguarded moment as they hang drowsily from a limb.

Collecting small mammals offers many thrills, some of which rival those of the big game hunter. C. Hart Merriam (1884) states that though he has been fortunate enough to shoot fourteen hoary bats, he would rather kill another than slay a dozen deer.

The hoary bat is usually not a social creature, and the males undoubtedly are solitary. Females have been observed hunting together, and five individuals have been seen within a short time over a small body of water. During the day, hoary bats hang among the foliage.

Usually, two young are born, although the four mammae will accommodate more. Birth usually occurs from mid-May through mid-June. The young weigh about 4.5 grams and have a forearm length of about 18 or 19 mm at birth. It is known that they grow very rapidly, attaining sufficient size and strength within a month to fend for themselves.

Most naturalists believe this to be a resident of the boreal region during the summer, but the species is known to breed at least as far south as Indiana and Pennsylvania. It does not appear that the young are carried with the mother on her hunting sojourns, but they attain a good size before being weaned. A female with two young attached was found on the ground in Denver. Examination showed the combined weight of the two young males to be more than 25 percent more than that of the mother.

The hoary bat, because of its great size, can conquer large nocturnal insects, but it does not disdain small prey. The stomach of a Pennsylvania bat collected by Earl Poole (1932) contained a large stink bug and a mosquito. Moths were the dominant food of three individuals from Indiana, although chrysomelid beetles and muscoid flies were also present. Anthony Ross found New Mexico hoary bats also feeding heavily on Lepidoptera, but a variety of other insect foods were eaten.

The hoary bat is a hardy creature and remains active in northern latitudes long after other bats have departed for the south or have retired to caves for the winter. It is commonly believed that hoary bats make a pronounced southerly migration during October and November, and indeed, a number of large migratory flocks have been reported. However, the southern area where numbers of this species spend the winter is not known. It is clear that some individuals remain in the north throughout the winter. At this season they are more often seen about human habitations. One has been taken from beneath a piece of driftwood on a Long Island beach. A hoary bat was seen flying about in a Pennsylvania forest in mid-day during a February thaw, and two have been taken in midwinter in Indiana. One of these was particularly interesting. It was taken on the side of a brick building at Terre Haute, on January 31, 1967 after a four-day warm spell. Maximum temperatures in the previous four days ranged from 58 to 67°F. The bat was in excellent condition and weighed 18.9 grams. The stomach was about a third full, containing grass, leaves, and shed snake skin. The small intestine was empty except for twenty-five trematodes. The large intestine contained a mass of greenish vegetation., including grass leaves. We believe this bat had stopped insect food in the fall and had eaten vegetarian prior to hibernation. It then voided the stomach and small intestine, leaving the vegetation in the large intestine as a fecal plug. It entered hibernation but emerged during the warm weather. With the return of cold weather, the bat again fed upon grass leaves and, by chance, also snake skin. It seems more likely that some (or many?) individuals hibernate rather than migrate as is commonly thought for this species.

This and other eastern vespertilionid species make an audible chatter during flight.

Northern Yellow Bat. *Lasiurus intermedius* (H. Allen)

DESCRIPTION.   Somewhat similar in appearance to the red bat but with different ears; color yellowish brown. The ear is of medium height, rather broad and rounded, only sparsely haired on inner surface, furred above halfway on outer surface; the tragus is broad basally, tapering at the tip, interfemoral membrane well haired above for about the basal half, naked below; there is a sprinkling of fur on underside of volar membranes along forearm to wrist; pelage is long and silky. Three adults from Florida average: total length, 115 mm; tail, 51 mm; hind foot, 10.2 mm; wingspread, 300–310 mm (Figure 3.23).

DISTRIBUTION.   The yellow bat has been taken at the following Florida localities: Lake Geneva, Clay County; Gainesville, Alachua County; St. Mary's River near Boulogne, Nassau County; Seven Oaks, Pinellas County; Lakeland, Polk County; Old Town, Dixie County; Davenport, Polk County; and Lake Kissimmee, Osceola County. It is not uncommon in the delta region of Louisiana and has been collected in Hancock County, Mississippi. The yellow bat occurs in the southern parts of Louisiana, Mississippi, Alabama, and Georgia, in all of Florida, and north to southeastern South Carolina (Figure 3.24).

HABITS.   Little is known of the yellow bat's habits. It is not common in collections and it apparently does not resort to caves or buildings as many other Chiroptera do. This species is closely associated with Spanish moss, and indeed, its range closely approximates that of Spanish moss. It is a permanent resident throughout its range, and is often the most abundant bat where it occurs.

Figure 3.23. Head of northern yellow bat, *Lasiurus intermedius*. Photo by Roger Barbour.

Yellow bats usually forage fifteen to thirty feet above the ground in open or scrubby areas. Feeding aggregations are formed after the young begin to fly.

Little is known of reproduction, although males appear reproductively active from July through mid-February. Three or four young per litter are produced in late May or June. Newborn young weigh about 3 grams and have a forearm length of about 16 mm. There is no evidence that the young are carried with the mother during feeding flight, but they do remain with her when she is disturbed from a diurnal roost.

The stomach of a yellow bat collected by H. B. Sherman (1939) at Gainesville, Florida, contained fragmentary insect remains which included Homoptera, Zygoptera, Diptera (Anthomyiidae), Coleoptera (Dytiscidae and Scolytidae), and winged myrmicid ants.

Evening Bat. *Nycticeius humeralis* (Rafinesque)

DESCRIPTION. *Nycticeius* can at once be distinguished from *Myotis*, which it somewhat resembles, by its short, sparse, dull brown fur and the

Figure 3.24. Distribution of the northern yellow bat, *Lasiurus intermedius*.

dentition. There is only one upper incisor on each side, and four molariform teeth behind the canine, none reduced in size. *Myotis* has two upper incisors per side, and the first two of the six molariform teeth are reduced in size, creating the appearance of a space between canines and large molariforms (Figure 3.1). The fur is dull amber or mummy brown above; the bases of these hairs are plumbeous. The ventral fur is paler. The young are considerably darker than the adults. The fur is closely confined to the body, not extending on the wing or tail membranes. The small ears are thick and leathery. The short, broad and low skull has a nearly straight dorsal profile. Evening bats resemble a small *Eptesicus,* but the short forearm (about 36 mm) and single upper incisor distinguish it from that species.

Average measurements of thirty adults from various parts of the range are: total length, 92.7 mm; tail, 36.8 mm; hind foot, 7.1 mm; forearm, 35.6 mm; weight, 5–6 grams.

Figure 3.25. Distribution of the evening bat, *Nycticeius humeralis.*

DISTRIBUTION.   This is a southern species. It occurs as a summer resident sparingly from Pennsylvania to southern Michigan and Illinois but reaches its maximum abundance in the south (Figure 3.25).

HABITS.   The evening bat leaves its roost in a hollow tree or building as darkness falls, spiralling from a height of forty-five to seventy-five feet until it is low over the ground. It has a slow and steady flight. Specimens have been secured by swishing a reed fishing pole in their path. This species almost never enters caves, although sometimes individuals are found among bat swarms at cave entrances in late summer.

As with other bats, the sexes segregate during the season of parturition, and few males are captured at this season. During August these bats appear unusually numerous, and at this time both sexes are found together, probably for breeding purposes.

The two young are born in late May to mid-June. Females with embryos have been taken as far north as southern Michigan. The nursing females leave their young in the recesses of the hollow tree or old building which serves as home as they forage for food. The forearm is about 14 mm at birth. Four stomachs of evening bats from Indiana have been examined, two by Anthony Ross and two by Whitaker. One of those examined by Ross contained beetles and flies, including Scarabaeidae and Drosophilidae; the other was nearly empty but contained one flying ant and one cercopid. One of the bats examined by us contained Delphacidae (Homoptera) 35 percent, Coreidae (Hemiptera) 20 percent, Scarabaeidae (Coleoptera) 35 percent, and unidentified matter 10 percent, the other contained 50 percent moths, 30 percent carabid beetle, and 20 percent Hemiptera.

Townsend's Big-Eared Bat. *Plecotus townsendii* (Cooper)

DESCRIPTION.   This bat is at once differentiated from all other species, except for Rafinesque's big-eared bat, by the tremendous ears, which, when laid back, reach to the middle of the body. Peculiar glandular masses rise high on the muzzle well above the nostrils. The color above is clove-brown, shading imperceptibly into slaty gray at the base of the hairs. The fur of *Plecotus* is much softer, longer, and woollier than that of *Myotis*. The inner margins of the long ears are scantily furred at their bases. The wing and tail membranes are without fur. This species can usually be separated from Rafinesque's big-eared bat by its buffy underparts; the hairs of the dorsum are sharply bicolored, and the first incisor has no accessory cusp. In *P. rafinesquii*, the underparts are washed with white, the dorsal hairs grade from the brownish tips to the slaty bases, and there is usually an accessory cusp on the first incisor.

Average measurements of eleven adults from Pendelton County, West Virginia, are: total length, 99.8 (96–110) mm; tail, 45.8 (42–52) mm; hind foot, 11.5 (11–12) mm; forearm, 44.4 (42–47) mm; weight, 9–12 grams; wingspread, 291–319 mm.

DISTRIBUTION.   *Plecotus townsendii virginiana* is recognized as an eastern subspecies of the western big-eared bat, but there was much confusion between the two eastern species of *Plecotus*. *Plecotus t. virginiana* is found primarily in West Virginia and Virginia caves over 4,000 feet altitude, but has also been taken in Kentucky (Figure 3.26).

HABITS.   These peculiar bats excite much wonder in those who first make their acquaintance. In the east this is a cave species, and specimens are usually met with in the twilight zone, where they hang suspended from the walls or ceilings, their long ears spirally coiled and flattened against the neck. Direct a beam of light on them and they are immediately alert, the big ears twisting alternately as if to catch the minutest sound. These are among the wariest of bats, and take alarm at the least disturbance. *Plecotus* appears to be far more sensitive than other bats. While exploring a West Virginia cave in August 1931, Hamilton found a group of these bats. He wished to examine the entire lot, but they were far too high to collect by hand, so a shot was fired into the cluster. All dropped, but several had not been injured in any manner. All were males with greatly enlarged testes, which suggests that possibly the bats mate at this season. Later the bats were watched while leaving this cave during the late dusk. The opening was situated several hundred feet above the valley floor, and the emerging bats first flew yet higher, soaring and circling at a height of several hundred feet until it was too dark for successful shooting. They then descended until within a few feet of the ground.

This bat emerges late in the evening. It is seldom taken in mistnets, perhaps because it has a very sensitive echolocation system. Many bats go into daily torpor, which serves to conserve energy; however, this species does not. In eastern populations, hibernating individuals in caves are usually in small groups, although they occasionally number up to a thousand or more. The same caves may harbor both summer and winter populations in this species. There are a few caves in eastern Kentucky where both it and *P. townsendii* occupy the same caves in winter, but this is not known to occur elsewhere in the east.

Nursery colonies are in caves. The single young is born during late June, the mother carrying it about until it has become too heavy. It then hangs alone in the cave or other shelter until it is strong enough to join the parent on the nocturnal feeding jaunts. The young weigh about 2.8 grams at birth.

These bats appear never to congregate in large masses, seeming to prefer the association of a few to many.

Figure 3.26. Distribution of Townsend's big-eared bat, *Plecotus townsendii.*

All that we have examined for a clue to feeding habits contained only the remains of Lepidoptera. This species serves as host to large parasitic flies, *Trichobius corynorhini* (Streblidae) which are often obvious as they crawl about during hibernation of the bat.

Because of its intolerance to disturbance, and its scattered eastern populations, continued efforts are necessary to protect this bat and the caves in which

it lives. Indeed, it is currently proposed as an addition to the federal endangered and threatened list.

Rafinesque's Big-Eared Bat. *Plecotus rafinesquii* (Lesson)

DESCRIPTION.   This species may at once be recognized from all eastern
bats except *P. townsendii*, by the tremendous ears, which are more than 30
mm high and joined at the base. A thick wart-like enlargement between the
eyes and the nostrils gives it the common name of ''lump-nose.'' This species
is about the size of the big brown bat. Measurements of four adults from
Tennessee and Georgia average: total length, 99.5 mm; tail, 48 mm; hind
foot, 40.2 mm; weight, 7–9 grams (Figure 3.27).

Plecotus rafinesquii is distinguished from *P. townsendii* by its underparts
which are washed with white. The dorsal hairs show no marked contrast
between the dark brown or blackish bases and the pale brown tips. *The inner
upper incisor is bicuspidate.*

DISTRIBUTION.   Rafinesque's bat ranges widely over the southern states,
from North Carolina and Kentucky south to the Gulf of Mexico (Figure 3.28).

HABITS.   The habits of this species are not well known, and would repay
detailed study. However, this species appears to differ most noticeably from
P. townsendii in that it is often a species of the hollow trees or buildings of

Figure 3.27.   Rafinesque's big-eared bat, *Plecotus rafinesquii.*

wooded areas, for the most part, rather than being strictly a cave species. Summer colonies are most often found in buildings. This species will hibernate in caves in the northern parts of its range, and in Kentucky there are some populations that live in caves year-round. In southern Illinois it sometimes hibernates in silica mines. Otherwise, the habits appear to be quite similar to those of *P. townsendii*. When at rest, the long ears are coiled spiral-fashion about the neck, suggesting the horns of a ram. The species emerges late and

Figure 3.28. Distribution of Rafinesque's big-eared bat, *Plecotus rafinesquii.*

food is probably in great part adult moths. Breeding occurs in fall or winter and one young per year is produced.

Brazilian Free-Tailed Bat. *Tadarida brasiliensis* (I. Geof. St. Hilaire)

DESCRIPTION. *Tadarida brasiliensis* is the only regularly occurring eastern bat with a free tail (the tail extending for some distance beyond the interfemoral membrane). The upper lips are characterized by deep vertical grooves or wrinkles. There is only one upper incisor per side. In addition, the ears are short and wide and the fur is unusually short and velvety. The fur is a uniform warm brown above with a very brief light-colored area at the base of the hairs; yellowish below. The membranes are brownish. Average measurements of twenty adults from Florida, Alabama, and Louisiana are: total length, 91.9 (88–98.6) mm; tail, 33.2 (26.8–37) mm; hind foot, 8.6 (7.4–9.2) mm; forearm, 43.1 (41.5–45.5) mm; wingspread, 343–359 mm (Figure 3.29).

Figure 3.29. Brazilian free-tailed bat, *Tadarida brasiliensis.*

DISTRIBUTION.   This *Tadarida* occurs in the south Atlantic States and Gulf Coast, from Columbia, South Carolina, through Georgia, Florida, Alabama, and Louisiana. It is the commonest bat in Louisiana (Figure 3.30).

HABITS.   The free-tailed bat often infests houses, stables, and business establishments by the thousands. There are estimates of 50,000 or more in

Figure 3.30.   Distribution of the free-tailed bat, *Tadarida brasiliensis.*

single buildings in summer. Winter aggregations are much smaller. One building in Gainesville that contained about 10,000 free-tailed bats in summer had only a few hundred hibernating there in winter. Hundreds or even thousands leave their daytime retreat at dusk, flying considerable distances to a favorite feeding ground and not returning to their roosting places until dawn. These bats possess a penetrating musky odor which can be detected at some distance from the site in which a large colony is established. This species seems to avoid caves in the east, but several million free-tailed bats occupy the great Carlsbad Cavern of New Mexico. They have occupied this cave for centuries, and at one time the accumulated droppings covered an area a hundred feet or more in width, a quarter of a mile in length, and up to a hundred feet in depth.

These bats are the most rapid flyers of all North American bats, suggesting swifts in their mode of flight. The flight is singularly erratic. They take flight early in the evening, often in the afternoon. H. B. Sherman (1937) has studied the breeding habits of this species. In Florida mating occurs from mid-February to late March in various years, at about the time of ovulation. This bat is unlike many others in that the sperm are not stored in the uterus of the female for a considerable period prior to ovulation. Sherman believes the gestation period to be from eleven to twelve weeks. The single young is born from late May to late June, depending somewhat upon the year. The parent scratches the amnion to shreds, and, as the interfemoral membrane is so short and the wing membranes are not employed to hold the newborn young, it hangs by the cord until it can clamber to the breast. The placenta is not removed by the mother as happens with many vespertilionids. The young is not carried by the parent as it leaves the roosting site. The female will allow any young in the colony to nurse, rather than accepting only her own. The females become sexually mature at the age of about nine months. The male has a short sexual season, from February to mid-April, when spermatozoa are present in the testes and epididymis, but they are apparently incapable of breeding at other seasons. Sherman has likewise reported on the food of this species. Among the chief insect items taken from the stomachs of eight specimens, winged ants, chalcids, dytiscid beetles, chironomid midges, and small lepidopterans were most prominent. Photographic studies by H. E. Edgerton and associates have shown that the tail membrane, although very short, may be extended in flight as an aid in catching insects.

It has been long known that many species of bats can transmit rabies by bite, but D. G. Constantine (1962) demonstrated transmission via the air upon prolonged exposure. Several species of bats have been found to carry rabies apparently often without dying themselves. Bats do not show "furious" rabies as occurs in other animals. In Indiana, red, hoary, and big brown bats are most often infected.

Wagner's Mastiff Bat. *Eumops glaucinus* (Wagner)

The free-tailed *Eumops* appears somewhat like *Tadarida,* but its much greater size and peculiarly fierce appearance are sufficient to distinguish it from the latter. This species was first recorded from the United States at Miami, Florida, by Thomas Barbour (1936) who suggests that it may have come from Cuba via a fruit steamer. A specimen from Havana, Cuba, preserved in alcohol in the Cornell University Museum has the following measurements: total length, 130 mm; tail, 47 mm; hind foot, 11 mm; forearm, 59 mm. Flying distance over the Florida Straits from Cuba to the Florida keys is less than 100 miles, no difficult achievement for this strong flying bat. Charles English, a biology teacher at Miami High School, wrote Hamilton that five specimens of *Eumops* were collected in 1950 in the Miami area. A number of additional records have been noted in recent years.

4

Edentata

(Armadillos)

Edentates are primitive mammals, the teeth of which may be lacking or numerous; when present these teeth are conical or peglike, rootless, and deficient in enamel. The family Dasypodidae (armadillos) is further characterized by the ossified skin, forming a shell completely over the dorsal surface of the animal. The teeth are numerous, unrooted, and homodont in nature, and the forefeet are provided with strong, curved claws for digging. One species occurs in the United States, although the order has a wide range through much of Central and South America.

Nine-Banded Armadillo. *Dasypus novemcinctus* Linnaeus

DESCRIPTION. This mammal, the only representative of the edentates occurring in the United States, can be confused with no other species. It is best characterized by the shell-like, scaly skin with nine transversely joined bands over the back. The simple, peglike teeth, the long, apparently segmented tail, and the prominent claws likewise help to distinguish this queer animal. Soft-haired skin covers the belly and inner surfaces of the limbs. The type, an adult male, measures as follows: total length, 800 mm; tail, 370 mm; hind foot, 100 mm; weight, 9–15 pounds (Figure 4.1).

DISTRIBUTION. In the early years of the present century, the armadillo was restricted to south and central Texas, but it has moved northeast and appeared in Louisiana about 1925. Armadillos are now found in Mississippi, Alabama, and Florida (Figure 4.2).

Figure 4.1. Nine-banded armadillo, *Dasypus novemcinctus.*

HABITS.   Where it has invaded the southern states, the armadillo is found in brushy or waste lands, in areas where the soil permits easy digging and supports enough food for its needs.

E. W. Nelson (1930), from his long acquaintance with this animal in Mexico, wrote of its habits as follows:

As might be surmised from its appearance, the armadillo is a stupid animal, living a monotonous life of restricted activities. Its sight and hearing are poor, and the armoured skin gives it a stiff-legged gait and immobile body. From these charac-

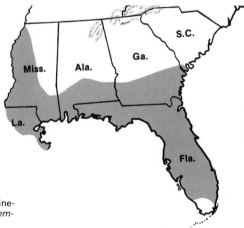

Figure 4.2. Distribution of the nine-banded armadillo, *Dasypus novemcinctus.*

teristics, combined with the small head hung on a short neck, it has in life an odd resemblance in both form and motion to a small pig; it jogs along in its trails or from one feeding place to another with the same little stiff trotting gait and self-centered air. If alarmed it will break into a clumsy gallop, but moves so slowly that it may be overtaken by a man on foot. So poor is its eyesight that a person may approach openly within about thirty yards before being noticed.

This species is an accomplished digger, and makes prominent burrows, from which trails radiate in several directions through the scrubby thickets. The clumsy gallop mentioned by Nelson is surprisingly rapid. If overtaken by man or hungry predator, it does not roll into a tight ball, as often claimed, although it will partially curl up, protecting the soft vulnerable belly from attack. More often it will run rapidly away, traveling with considerable speed for such a clumsy-appearing beast.

The food of the armadillo consists chiefly of beetles and their larvae, and other insects, which it secures by first exposing them with its sharp claws and then flicking them into the mouth with the long sticky tongue. Plant food makes up about 10 percent of the diet, and snails, slugs, millipedes, and centipedes are also important. Only two or three percent consists of birds' eggs and other controversial items. The scats of this beast are round mud balls, in the matrix of which are usually found numerous insect remains.

The nest consists of half a bushel or so of leaves or grass and is in a burrow. In loose soil an armadillo can dig itself out of sight in a few minutes.

The studies of H. H. Newman (1909) have shown that this interesting mammal almost invariably produces four young at a litter, always of the same sex. The early embryo divides by fission to form twin embryos, and immediately these twins divide to give rise to two pairs of duplicate twins or quadruplets. Thus they are all derived from one fertilized egg and inherit but one assortment of genes. Fertilization is delayed. The young are born in an advanced condition; the armour is soft and pliable at birth, but soon hardens. Mating occurs in summer, but implantation is delayed into November, when four months will elapse before the young are born. One litter a year is the rule.

Many armadillos are killed and their armored skin prepared into baskets for the tourist and curio trade. Mortality on the highway probably accounts for more deaths than predation. In spite of such persecution they seem to thrive, as the amazing extension of their range within recent years would indicate.

5

Lagomorpha

(Hares and Rabbits)

Lagomorphs are superficially like the rodents, with which they have much in common. There is, however, enough difference between them and the rodents to justify regarding each as ordinally distinct, the similarity being accounted for through parallel development or convergence.

Lagomorphs possess two pairs of upper incisors. The first pair is rodent-like in character, with a broad groove on the front surface. The second pair, placed directly behind the large first, is small, lacking the cutting edge, and is nearly circular in outline (Figure 5.1). The distance between the cheek teeth of

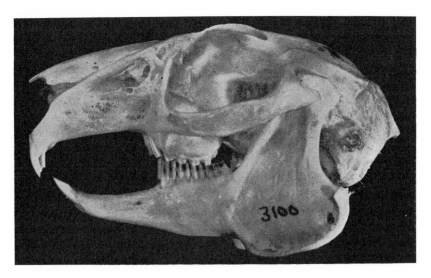

Figure 5.1. Skull of snowshoe hare, *Lepus americanus.*

115

the lower jaws is considerably less than that between the molar tooth rows of the upper jaw. As a consequence, only one molar row of the upper and the lower jaw are capable of opposition at the same time, resulting in an ectalental or sidewise movement. The fibula is a long strong bone, articulating distally with the calcaneum, a condition not found among the rodents.

Lagomorphs are cosmopolitan creatures, indigenous everywhere but in Madagascar, Australia, and New Zealand. Man has introduced rabbits into Australia, with dire results.

The lagomorphs are important mammals, for they supply felt and hides to the hatter and furrier. Millions are killed for sport and food. They often cause colossal damage to agriculture, especially in orchards, hay fields, or gardens.

Key to the Genera of Lagomorphs

A. Interparietals fused with parietals. Hind foot more than 100 mm . . . .*Lepus*
AA. Interparietals distinct. Hind foot less than 100 mm . . . .*Sylvilagus*

Key to Rabbits of the Genus Sylvilagus

A. Anterior extension of supraorbital process present. Posterior portion free of braincase or separated by a slit from braincase. Nape rich cinnamon . . . .
. . . .*Sylvilagus floridanus*
AA. Anterior extention of supraorbital process absent (or if a point is present, then at least five-sixths of posterior extension is fused to braincase).
    B. Basilar length of skull more than 63 mm . . . .*S. aquaticus*
    BB. Basilar length of skull less than 63 mm.
        C. Underside of tail white. Posterior extension of supraorbital process tapering to a slender point, this point free of braincase or barely touching it and leaving a slit . . . .*S. transitionalis*
        CC. Underside of tail brown or gray. Posterior extension of supraorbital process fused to skull, usually for entire length . . . .*S. palustris*

Marsh Rabbit. *Sylvilagus palustris* (Bachman)

DESCRIPTION.   The marsh rabbit is small and dark brown, with small, slender, dark reddish-buffy feet, and dingy (rarely white) color on the underside of tail; ears short and broad; tail very small. In color it is similar to *S. aquaticus,* but can be recognized by the smaller size and the dark underparts of the tail. Upperparts reddish brown; nape dark cinnamon rufous; rump, upper side of tail and back of hind legs varying from chestnut brown to dark rusty reddish; middle of abdomen white, the rest of belly buffy to light brown. Average measurements of seven adults from Florida and Georgia are: total length, 441 mm; tail, 34.5 mm; hind foot, 92 mm; weight, 3.5 pounds (1.5 kg) (Figure 5.2). Individuals from peninsular Florida are smaller, darker, and more reddish brown than typical *palustris;* upperparts dark reddish brown, shading on flanks and sides of abdomen into dark, slightly brownish buffy; sides of head and ears slightly paler than back and more grayish; often a buffy band across the middle divides the white belly into two irregular patches. Tops of fore and hind feet are rich dark cinnamon rufous. Average measurements of eight adults are: total length, 429 mm; tail, 42 mm; hind foot, 89 mm; weight, 2.2–2.6 pounds (1–1.06 kg).

DISTRIBUTION.   The marsh rabbit occurs from Dismal Swamp, Virginia, south through southern Georgia and all of Florida, and west to Mobile Bay, Alabama (Figure 5.3).

HABITS.   The slender-footed little marsh rabbits inhabit the brakes, wet bottomlands, and brackish swamps and sea islands along the Atlantic Coast, and dwell in the dense hammocks or the borders of freshwater lakes of Georgia and Florida. There is plentiful evidence of this rabbit on the sandy thickets lying back of the mangroves of south Florida beaches.

The marsh rabbit enters the water voluntarily and swims efficiently for considerable distances. Thus Ivan Tomkins records a specimen swimming strongly 700 yards from shore. Inasmuch as their range includes extensive areas of lowland, bordering tidal waters which periodically inundate their homes, these rabbits must be able swimmers. They do not appear to be particularly modified for an aquatic life, although the hind limbs are less furred and the nails somewhat longer than with cottontails.

Figure 5.2. Marsh rabbit, *Sylvilagus palustris.*

Figure 5.3. Distribution of the marsh rabbit, *Sylvilagus palustris.*

The marsh rabbits have well beaten trails in the brakes and dense marsh vegetation. While they can and do run like the cottontails, the marsh rabbits are likewise capable of stepping alternately with each foot, much in the manner of a dog or a cat. Their slow gait when feeding is thus quite contrary to the hopping motions of other rabbits.

Marsh rabbits are quite nocturnal over much of their range, although variable tide levels may make them move to some extent during the day. Hamilton watched an immature Florida individual for several days one spring and noted that it invariably returned to the same resting quarters under a brush pile each morning. These rabbits are usually common throughout their range, and although they may be well hidden during the day, a strong light will reveal numbers an hour or so after dusk.

Breeding commences in early February. Gestation is about thirty-nine to forty days. The two to four young (rarely five) are born in a warm fur-lined nest which occupies a rather sizable depression. They are born with the pelage well developed but the eyes closed. They remain in the nest for several days after they are weaned. They return to the shelter for a few days after they have first ventured out. Hamilton found one nest under a cabbage palm in Lee County, Florida, large enough to insert his foot into. Probably several litters are produced in a season; young and half-grown young have been found in October.

These rabbits eat a variety of plants and aquatic emergents, including cane, marsh grass, leaves of deciduous trees, and twigs. In addition, they dig up the rhizomes of amaryllis and the bulbs of several plants, including the wild potato, *Apios*.

This rabbit will thump the ground with the hind feet as is the case with many rabbits, and one was heard to scream when wounded.

The marsh rabbit seldom exerts economic pressure on crops. It is widely hunted in the south, where the dried grasses are burned over to rout it into the open where it can be clubbed or shot.

Eastern Cottontail. *Sylvilagus floridanus* (Allen)

DESCRIPTION.   In the coastal lowlands of peninsular Florida, the cottontail is a small, dark rabbit, varying in color from dark grayish-buffy to rusty, buffy brown; nape and legs rich cinnamon rufous; ears short, rounded and darker than back. The top of the head and back are dark buffy brown interspersed with reddish and dark buffy hairs; rump and sides of body dark buffy gray, washed with black; top of tail dull rusty brown; front of forelegs deep cinnamon rufous; underside of neck dull dark ochraceous buff, lower flanks clearer buff; outside of ears dark grayish buffy, heavily bordered and washed with black, especially the terminal half. The rusty color is lost in the summer, being replaced by grayish buffy brown. Average measurements of five Florida specimens are: total length, 436 mm; tail, 45 mm; hind foot, 90 mm (Figure 5.4).

Cottontails from the southeast are larger and paler than those from Florida, with larger ears and prominent gray rump patch; top of head and back dull, rather dark rusty yellowish or slightly rusty ochraceous buffy, paler and less heavily washed with black; nape rich rusty rufous; outside of ears dull grayish buffy; front and outside of forelegs dark rusty rufous; tops of hind feet whitish or pale rusty buffy. Average measurements of five North Carolina specimens are: total length, 446 mm; tail, 65 mm; hind foot, 94 mm; weight, 3–3.4 pounds (1360–1430 g). Animals from the northern part of our area are also large, but with shorter ears and longer hind feet; top of head and back pale pinkish buffy, darkened by the overlying wash of black; sides of head and body grayer than back and usually much paler, with fewer black-tipped hairs; gray rump patch as in those from the southeast but washed and darkened with black. Average measurements of ten New York specimens are: total length, 414 mm; tail, 56 mm; hind foot, 101 mm; weight, 2.7–3.2 pounds (1.2 kg). Cottontails from the lower Mississippi River valley are small, and best characterized by the reddish cast of the head and dorsum. Top of head and back deep

Figure 5.4. Eastern cottontail, *Sylvilagus floridanus.*

ochraceous buff, strongly washed with black, giving a rusty or reddish brown effect, sides of body paler than back; grayish rump patch poorly marked; top of tail reddish brown; front and sides of forelegs rich deep ferruginous; underside of neck deep buff varying to dull dark ochraceous buff. Average measurements of five adults are: total length, 418 mm; tail, 56 mm; hind foot, 92 mm.

Animals from extreme southern Florida are smaller, paler (grayer and less washed with black) than northern Florida cottontails.

DISTRIBUTION.   The eastern cottontail occurs throughout eastern United States except for northern New York, and northern New England (Figure 5.5). The cottontail is moving into the southern Adirondacks. It has probably been introduced into New Hampshire.

HABITS.   Over its wide distribution, the cottontail occupies diverse habitats, from swampy woods to upland thickets and farmlands, and residential areas of sizable cities. It seldom occurs in any numbers in the heavy forest, but is quick to take advantage of new territory opened up by lumbering.

Cottontails are timid beasts, relying only on their speed to escape a world of enemies. During the day they remain in a form; either at the base of a suitable tree, the scant shelter of a grass tussock, or under the coverage of a blackberry thicket. Here the cottontail sleeps away the day, ever alert to explode into flight if threatened by a marauding fox or other enemy. In the

Figure 5.5. Distribution of the eastern cottontail, *Sylvilagus floridanus*.

northern part of its range, this species often utilizes the burrow of the wood-chuck, skunk, or some other creature during the colder winter months. It rarely, if ever, prepares its own burrow.

Cottontails seldom roam over an area greater than a few acres. The winter snow reveals their trails, crisscrossing through a sumac patch or bit of swamp. The tracks are often so numerous that one obtains a false notion of their abundance. Cottontails are active chiefly at night, moving from their forms with the approach of dusk.

During the summer these rabbits eat a wide variety of green foods, chiefly grasses and the low broad-leafed weeds. They may become a nuisance in the garden, leveling the rows of young peas, beans, and early summer crops. Winter foods consist primarily of the buds and tender twigs of many small trees and bushes. Sumac bark is particularly favored. The canes of blackberry and other thorny stalks are neatly cut, and sapling sprouts are severed at the snow line. Like many other lagomorphs, this species practices coprophagy or reingestion. Green food is rapidly swallowed and the rabbit returns to protective cover where soft green pellets are defecated. These consist of undigested vegetation and are leisurely eaten.

The cottontail is a prolific species; several litters usually of three to six (extremes one to seven) young are produced over a long breeding season. Gestation is twenty-eight to thirty-two days. Breeding may commence in early January, but the young are not commonly born until March. The parent makes a warm nest by scraping out a shallow depression, or utilizing a natural cavity, which is lined with finely shredded leaves and grasses, and with a good amount of fur which she removes from her belly and breast. This fur adds substance and warmth to the natal chamber. The nest is carefully covered when the mother is away, which is usually during the daytime. The nest may be almost any place, even in mowed lawns of residential areas. The young are blind, essentially naked and quite helpless at birth, weighing slightly less than an ounce (28.3 g). The mother returns to the nest each dawn and dusk for about sixteen days, opens the top, and lies over it to let the youngsters nurse. After the young are licked clean, the nest is carefully closed, and the mother feeds and rests nearby. A gardener can weed and hoe within inches of the camouflaged nest without knowledge of its presence.

Rabbits have many enemies. Scores of young are routed from their nest cavity by wandering dogs, foxes, skunks, and crows; the growing youngsters are preyed upon by hawks and several species of owls, the larger snakes and even the red squirrel; the adults also fall prey to most of these enemies. Highway mortality is severe, for the rabbits are partially spellbound by bright headlights and run ahead of the automobile in a dazed fashion.

The cottontail is the most important game animal of eastern United States. Millions are shot each season. The flesh is excellent; few meals compare with a well prepared bowl of rabbit stew. The pelt is of little value, but is used chiefly in trimmings on children's garments.

New England Cottontail. *Sylvilagus transitionalis* (Bangs)

DESCRIPTION.   A small short-eared cottontail with pinkish buff coat heavily washed with black. It may be distinguished from *S. floridanus* by its shorter ears and black patch between the ears. In addition, the anterior supraorbital process of the skull is quite distinct. In *transitionalis,* this process narrows conspicuously along its anterior outer side, resulting in the absence of the anterior process of the supraorbital, thus rendering the anterior notch obsolete or reducing it to a shallow concave depression. In *floridanus* the supraorbital is broad and heavy, nearly on a plane with the frontal area, giving a broad frontal area and adding to the heavy appearance of the skull (Figure 5.6). The postorbital process of *transitionalis* usually tapers to a point and does not touch the skull whereas in *floridanus* it is broad, being nearly the same width throughout its length and at its distal extremity fusing with the skull. Measurements of ten adults from various parts of the range average: total length, 420 mm; tail, 57 mm; hind foot, 97 mm (Figure 5.7).

DISTRIBUTION. *Sylvilagus transitionalis* occurs from southern Maine, northern Vermont, and western New York south in the Allegheny Mountains to southeastern Alabama (Figure 5.8).

HABITS.   The New England cottontail is a wood or brush species, preferring open woods or their borders or shrubby areas and thickets over more or less open areas. Its behavior does not differ materially from that of *Sylvilagus floridanus*.

During the summer months this cottontail eats herbaceous plants, especially clover, and succulent grasses, often feeding well after sunrise. As dusk approaches, it leaves its retreat in a thicket and is particularly active during the few hours following sunset. Reingestion occurs.

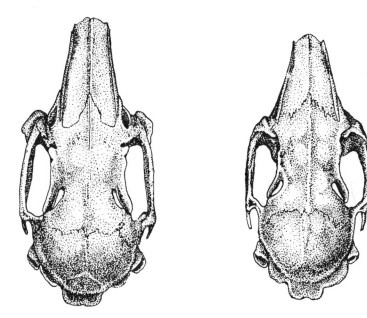

Figure 5.6. Skulls of *Sylvilagus floridanus* and *S. transitionalis.*

Figure 5.7. New England cottontail, *Sylvilagus transitionalis.*

Several litters are born from late winter to late summer. Hamilton collected a pregnant female during the latter part of August, and from the size of the embryos, these would probably not have been born until mid-September. Gestation is believed to be about twenty-eight days. Three to eight young per litter are produced with a mean of about 5.2. Although this rabbit lives in company with its larger relative, *Sylvilagus floridanus,* specimens are seldom taken which indicate that the two species interbreed.

As the great northern coniferous forests have been felled by the axe, the timid and retiring snowshoe hare has been pushed back and the new frontier of scrubby second growth invaded by the cottontail. There are some who hold that the cottontail, although smaller, can master the hare in a fight, and that its presence is sufficient to drive out the hare. Whether this reasoning is correct or not has yet to be demonstrated. It is probable that the modification of the habitat by man is the primary cause of this population change.

Some hunters differentiate between the smaller *transitionalis* and the larger, longer-eared *floridanus*. They call the former ''wood rabbit'' and the latter ''brush rabbit.'' *Transitionalis* will ''hole up'' when chased by dogs.

Swamp Rabbit. *Sylvilagus aquaticus* Bachman

DESCRIPTION.   This large, yellow-brown, big-headed, short-furred rabbit is readily distinguished from the cottontails by its short sleek fur, thin-haired

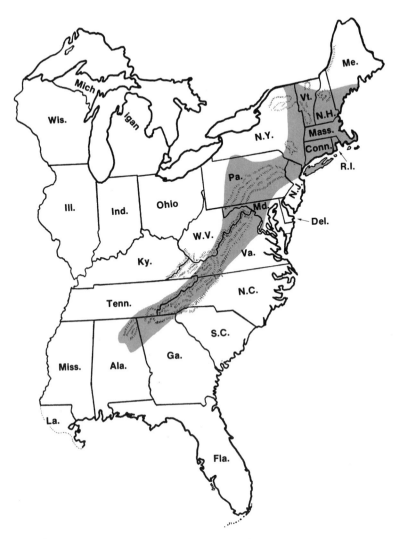

Figure 5.8. Distribution of the New England cottontail, *Sylvilagus transitionalis.*

slender tail and dark coloration. While it resembles the marsh rabbit, the much greater size serves to distinguish it from this related species, although the way the ranges abut without overlapping tempts one to hypothesize that the two might belong to the same species. The top of the head is buffy brown; back buffy grayish brown to rusty brown; rump and upper sides of legs rusty brown; sides of head and body paler than rest of back; upper sides of feet and legs cinnamon rufous; outsides of ears browner than sides of body; lower neck buffy gray; rest of underparts, including lower tail, pure white.

Average measurements of ten adults from Louisiana, Alabama, and Georgia are: total length, 522 mm; tail, 67.5 mm; hind foot, 105.1 mm; weight, 3.5–6 pounds (1.6–2.7 kg). Individuals from the narrow belt of swamps and marshes within the upper limits of tidewater along the Gulf Coast from extreme southern Louisiana, Mississippi, and Alabama are similar in size, but much darker and more reddish; upperparts dark rusty or reddish brown strongly washed with black, and becoming distinctly more rufous on lower rump and top of tail. In summer the black wash is lacking, the reddish color fades, and the rabbit becomes pale brown, in a measure resembling a cottontail; the shoulders are grayer (Figure 5.9).

DISTRIBUTION.   This rabbit occupies the southwestern portion of our area, from southern Illinois, extreme western Kentucky and Tennessee, south to the Gulf Coast, and east through Alabama, northern Georgia, and extreme western South Carolina (Figure 5.10).

HABITS.   The swamp rabbit chooses the wet bottomlands, the river swamps, and impenetrable jungles of cane. Throughout its range, this rabbit appears to favor the wet ground, although it moves to higher points when pursued by hounds. Spring floods periodically drive it to the levees and higher

Figure 5.9. Swamp rabbit, *Sylvilagus aquaticus.*

Figure 5.10. Distribution of the swamp rabbit, *Sylvilagus aquaticus.*

knolls, where hunters easily kill large numbers with clubs. In west Texas, Hamilton has seen it in mesquite thickets.

This species is a favorite with the hunter, but its speed is more than a match for that of any dog, and few would be taken if it were not for the unfailing habit of taking refuge in a hollow tree or other likely cavity after a short run. The great, slightly furred, splayed toes, with sharp nails, offer the swamp rabbit security in the softest mud, where the marks of its characteristic gait, often a walk rather than a hop, leaves mute testimony to its abundance.

Swamp rabbits are accomplished swimmers, striking out boldly across a sizable body of water when pursued, or visiting the small islets which dot their watery home. The rabbit will take to the water when alarmed, its nose and ears alone visible above water. In spite of many assertions to the contrary, all available evidence suggests that this rabbit is a more accomplished swimmer than its smaller cousin, the marsh rabbit. In some parts of its range, the latter animal ventures into the water only when pressed to the utmost, but the swamp rabbit swims for considerable distances, apparently as a regular means of moving into new quarters. Robert Kennicott (1859) states that it not only takes to the water and swims readily, but even dives without hesitation when pursued.

Little is known of the breeding habits. The few accounts state that two litters are produced in a year. The young number from one to four, averaging about two or three. In Louisiana they are born in any month of the year, peaking from February through May. The gestation period is thirty-nine to forty days. The nest, as for many other rabbits, is a slight depression lined with fur.

The swamp rabbit feeds upon a variety of herbs and succulent aquatic plants. Its fondness for the stems of the ubiquitous cane (*Arundinaria*) is well known to southern hunters, who beat these tangles to put up the "cane-cutter." Swamp rabbits commonly defecate on logs and stumps, thus giving a clue to hunter or biologist that this species is about.

A. H. Howell (1921) states that when cultivated fields adjoin the swamps, these rabbits often forage in corn or other crops and at times cause considerable damage. From the very nature of its chosen habitat, this rabbit seldom occasions any mischief to the farmer, and it is hunted for the sport and its delicious meat, rather than through any desire to reduce its numbers.

Key to Hares of the Genus Lepus

  A. Tail dark above and below . . . .*Lepus americanus*
AA. Tail white or white below.
    B. Tail white (or faint line dorsally that does not extend onto rump) . . . .
      . . . .*L. townsendii*
    BB. Tail white below, with black above.
      C. Hind foot dorsally white or whitish . . . .*L. californicus*
      CC. Hind foot dorsally without white . . . .*L. europaeus*

Snowshoe Hare. *Lepus americanus* Erxleben

DESCRIPTION. The snowshoe hare is a medium-sized hare, with very large and long hind legs admirably adapted for running and jumping; large ears, short tail, and dense fur; the soles of feet are well furred and this trait particularly pronounced in winter. Summer pelage: upperparts generally dusky grayish or grayish brown; top of head dusky yellowish brown; sides of head cinnamon to buffy; back dusky grayish brown, strongly washed with black along middle; black replaced by gray or yellowish brown on sides; rump more heavily washed with black; top of tail black, white below; tips of ears bordered with black; underside of neck variable, usually a buffy cinnamon; underside of head and middle of belly white. Winter pelage: tips of ears dusky, rest of pelage white (Figure 5.11).

Figure 5.11. Snowshoe hare, *Lepus americanus.* Photos
by Robert Green.

Average measurements of seven adults are: total length, 473 mm; tail, 42
mm; hind foot, 135 mm; weight, 3.5–4.5 pounds (1.6–2.04 kg). Individuals
from extreme eastern Maine have slightly longer ears and more cinnamon in
the pelage; populations of the Allegheny Mountain area are the largest and
most brightly colored. These animals are best characterized by the rusty
ochraceous brown upperparts which are sometimes strongly tinted with
olivaceous; underparts of neck and a line along flanks rufous, in sharp contrast
to upperparts. Individuals from Wisconsin and western upper Michigan are
paler. Summer pelage: The top of head and back is dull buffy, varying to pale
dull ochraceous buffy brown, darkest on head; and top of back only slightly
darker than sides of body; rump slightly more washed with black than back;
tip of tail mixed black and dingy white, lending a dusky grayish or buffy
color; sides of head, especially about eyes and back to base of ears, richer,
clearer, and more ochraceous buffy than back. Winter pelage is entirely pure
white except a well-marked blackish border about tips of ears; the ochraceous
buffy showing through strongly whenever the white-tipped fur is parted.
Weight is 3.2 to 4 pounds (1450–1815 g).

DISTRIBUTION.   The snowshoe hare is found in the eastern United States,
northern Wisconsin, much of Michigan, and New England and New York,
south through most of the Alleghenies. It is now rare in the Smoky Mountains
(Figure 5.12).

HABITS.   The varying hare is a forest species, never far from the dense
woods. It prefers the brushy semiopen tracts surrounded by the evergreen

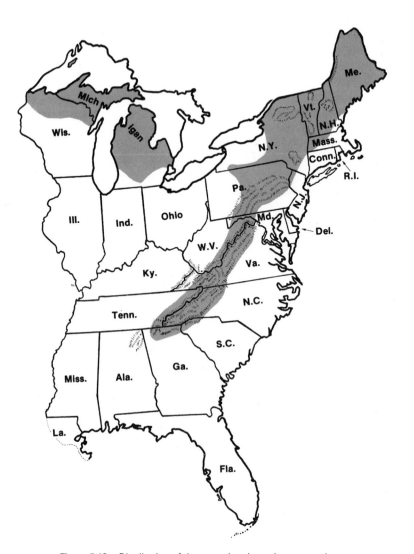

Figure 5.12. Distribution of the snowshoe hare, *Lepus americanus.*

forest, dense cedar swamps, and the sparsely wooded hillsides which support much brush. In the southern mountains the thickets of laurel and rhododendron are its home.

One of the characteristics of this species, as with many northern animals, is its great abundance in some years and almost total absence in others. This population change is best reflected in the fluctuating take of fur animals, such as the lynx and fox, which depend in large measure on the snowshoe hare for their existence.

During the day these large hares sit quietly in the form, forever alert to the many dangers hidden about. If alarmed, the hare leaps away, covering a dozen feet in a single bound, and circling through its home range of several acres. If not pursued, it soon comes to rest, the large ears raised and the twitching nose and big eyes alert to detect a new menace.

With the approach of the northern winter, the brown summer coat is shed and replaced by white. This is not a phenomenon occupying a day or two; it is a slow process covering two or three months; and a similar period is required for the vernal change back to brown. The white coat is unquestionably of great aid to the animal in concealing it from enemies; indeed, the very habits and actions of the animal are modified to the protective coat.

The big furry feet permit these hares to race with complete abandon across the snow where other animals flounder. They usually run in well marked trails, both in summer and winter. These highways become very pronounced because of the packed snow, and are often utilized by the trapper, who places his snares in them. The hare provides both meat for the trapper and bait for his cubby sets.

The hare appears to make no other home than its form beneath a low coniferous branch, a tangled mass of brush or an alder thicket. Occasionally it utilizes a hollow log, the cavity beneath a rotting stump, or even a woodchuck hole.

During the summer, succulent grasses and berries provide the hares with a sufficient and varied diet. In the winter the slim twigs and buds of the alder thickets, the tender balsam tips, and the shoots of many hardwoods and conifers offer coarse but palatable food, and these hares are seldom hungry. As successive snows add thickness to the winter blanket, the hares are forced to prune the higher branches, and at times may severely injure the bark and trunk by tearing the cambium with their strong grooved incisors.

Unlike true rabbits (*Sylvilagus*), the snowshoe hare gives birth to precocious young, which are covered with dense fluffy fur and have open eyes. So far as is known, no nest is prepared for the youngsters. The gestation period is thirty-six days (Wallace B. Grange). Probably two or three litters are produced from March to August, the litter size varying from one to six. Four young appear to be the usual number.

The snowshoe hare is beset by many enemies. Bobcats and foxes succeed in capturing large numbers, and the fisher, lynx, and other predators prey on this species. The goshawk and great horned owl are dreaded foes.

The snowshoe hare is a favored game animal. Its spirited runs and the fact that it never holes up when pursued permit much sport. In addition, its flesh is well flavored and in the far north the fleecy hide is stripped and sewn into warm robes.

Black-Tailed Jackrabbit. *Lepus californicus* Gray

DESCRIPTION.   A hare with very long ears and very large hind feet. The soles of the feet are densely furred. Upper lip is divided. The back is buffish gray, interspersed with blackish. The tail has a black stripe above, extending onto back, this stripe bordered by white on tail. The ears are brownish, with last 20–30 mm usually black. The animal is similarly colored summer and winter. Ranges of measurements are: total length 465–630 mm; tail 50–112 mm, hind foot 112–145 mm; weight 4–8 pounds (1814–3630 g).

DISTRIBUTION.   The black-tailed jackrabbit occurs naturally in the western United States, but was introduced in the Miami region of Florida in the 1930s through simply releasing them for greyhounds to pursue as a training technique. Many escaped and took up residence in the pastures and sand prairies west to the Everglades. They first became very abundant, then were nearly exterminated by flooding. However, they still persist and are fairly common locally. They are especially common around Miami International Airport.

This species was introduced on Nantucket Island, Massachusetts, from Kansas in 1925 solely for the purpose of sport. It is hunted after the manner of the English fox hunts. The jackrabbit has adapted to the arable land, stabilized dunes, and beach grass.

HABITS.   In the west this species is found in barren areas such as prairies, pastures, cultivated fields, and meadows, but also in areas of higher vegetation. In vegetation they often make trails. Black-tailed jackrabbits spend the day resting in forms; they apparently never use burrows. They dig the forms themselves and may use them once, over short periods, or over long periods. The young are born in a deeper form, which is sometimes lined with fur from the female's breast.

Black-tailed jacks are excellent and swift jumpers. They often move in 5 to 10 foot hops, but can leap 20 feet at a jump when in a hurry, going 2 to 5½ feet off the ground. Every fourth or fifth jump is exceptionally high, seemingly giving the animal a better view of its surroundings. The rabbits can attain speeds of 30 to 35 miles per hour for short periods. They are also good swimmers, taking readily to the water when frightened, but will swim even when not pressed. They shake when they emerge in the manner of dogs. The rabbits establish home ranges of less than 50 acres, with considerable overlap.

Most rabbits are generally silent, but can squeal when in distress, and this species is no exception. Other calls occur when fighting and also to bring the young together. This species has involved prenuptial behavior patterns, involving long chases, boxing, and leap-frog. Tufts of hairs are sometimes lost, possibly due to biting. Breeding may occur throughout the year in some localities. Gestation is about forty-one to forty-seven days, with a female producing one to four litters per year. There are one to eight young per litter. The young are precocious; nevertheless, they are born in a nest in a form, and the nests are covered when the female leaves. The female comes to the nest and nurses the young several times each night. Young in captivity were particularly active at night.

Fresh and herbaceous plants (including crops), are major food in summer, whereas woody and dried herbaceous vegetation are major foods in winter. As with many lagomorphs, food eaten rapidly in the field is swallowed where eaten, and forms green "soft pellets." These pellets are then defecated and eaten in the relative safety of the form, a process called reingestion or coprophagy. The soft pellets are gleaned directly from the anus and the rabbit swallows them whole thereby passing most food twice through the alimentary canal before the production of hard fibrous fecal pellets. Many plant species in an area are eaten, with feeding taking place mainly in open areas near weed patches. Feeding is most common in the evening, but may take place at any time, day or night.

Aggregations of jackrabbits sometimes occur, but it is thought these are due to exceptionally good food and cover conditions in local areas rather than to any developed social organizational qualities. They are alert creatures, with the ears immediately being raised at any indication of danger.

Black-tailed jacks, especially the young, are preyed upon by coyotes, foxes, hawks, owls, and snakes.

White-Tailed Jackrabbit. *Lepus townsendii* Bachman

DESCRIPTION.   This is a large hare, whose total length is more than 550 mm. The length of the hind feet is greater than 150 mm; the ear more than 90 mm; and the tail at least 65 mm. Summer pelage is pale buffy gray above. The tail is white above and below, or sometimes there is a narrow, dusky stripe dorsally on the tail which does not extend onto the body. The ears are buff or buffy gray on anterior half of outside, whitish with broad black patch extending to tip of ear on posterior half. Venter is white or pale grayish. Most Wisconsin individuals are pure white in winter. Measurements of Wisconsin individuals are: total length 575–670 mm; tail 70–115 mm; hind foot 148–175 mm; and weight 5.8–9.5 pounds (2620–4308 g).

DISTRIBUTION.   In the eastern United States the white-tailed jackrabbit occurs in northwestern Illinois and most of Wisconsin. It was probably introduced (Figure 5.13).

HABITS.   This is a very large hare, but it is seldom seen. It leaps away rather than heading for cover. It is a species of the prairie, and will inhabit open areas such as plowed fields, stubble, pastures, grasslands, barrens, and

Figure 5.13. Distribution of the white-tailed jackrabbit, *Lepus townsendii.*

burnt-over areas. White-tailed jackrabbits are active at dusk and dawn especially, or at any time during the night. They are seldom active during the day unless disturbed. When at rest they sit, often in a form, under a tussock, or in a furrow, or simply on the open ground, with ears folded back. When in this position, they are nearly invisible, with their gray-buff color of summer blending into the soil color, or the white of winter blending into the snow. Often the rabbit may be approached quite close before it will suddenly spring up and bound away, twelve to sixteen feet at a leap, and at speeds up to but seldom exceeding thirty-six miles per hour. This species will enter the water when cornered. It swims by paddling with the front feet and using the back feet in a leaping motion. Like other rabbits, this species is usually silent and is protected by remaining motionless and blending into the environment. However, like many rabbits, it has a call. It is a series of three or four short notes in quick succession. This call may be an alarm note. The large ears are moved about in various directions, sometimes one backward and the other forward, as the jackrabbit listens to the sounds around him.

Mating occurs in mid-April or later. The young are born after a thirty-day gestation period and are precocious. There is no nest for production of young. The young may be dropped on the ground, or in a form used for resting by both sexes. The young are well furred and their eyes are open. They are soon active, doing some foraging for themselves when about two weeks of age. Weaning takes place at about one month, and the young become entirely independent at about two months. There may be two litters per year in Wisconsin.

The form is rather simple, and may consist only of a slight depression, or it may be up to two feet long, a foot wide, and six inches deep. However, it averages probably two inches deep.

Summer food consists of various green vegetation, such as clover, grass, alfalfa, or green shoots and leaves of many species of wild and cultivated plants. In winter the rabbit browses on buds and twigs and on dry vegetation as it may find, such as hay or straw, or the leavings from the previous year's harvest.

This species is not colonial, and does not generally occur in great enough numbers to cause much harm. On the other hand, it is a fine game animal, both for the sport it provides, and for its meat.

European Hare. *Lepus europaeus* Pallas

DESCRIPTION.   Somewhat similar in appearance to our western jackrabbits, this large hare may be recognized by its five-inch ears and the thick mantle of somewhat kinky guard hairs (Figure 5.14). The jet upper part of the rather long tail is also distinctive. The hare weighs as much as 14 pounds (6.3 kg).

DISTRIBUTION.   A native of Europe, this great hare was introduced into Dutchess County, New York, in 1893. Importations continued through 1911, the purpose being to provide a game species. A gradual buildup of the population resulted in rather widespread but spotty distribution in the northeast.

Figure 5.14. European hare, *Lepus europaeus.* Drawing by Marjorie Crimmings McBride.

Additional introductions were made in Ontario, where the animal is now well established.

HABITS.   The European hare is a creature of open fields, shunning the forest or heavy brushland. The great hind legs allow the hare to run like a dog, occasionally standing on tiptoe. A speed of thirty-five miles per hour has been recorded.

Grass and herbs provide a summer diet, while the winter fare consists of twigs and buds and the bark of shrubs and small trees. The damage to orchards is often severe. In 1932, Hamilton inspected an apple orchard at Poughkeepsie, New York, where the hares had stripped the bark from the trunks of young fruit trees, leaving only shredded remains to indicate the destruction wrought by their nocturnal forays.

The two or three leverets are precocious at birth, covered with long silky grizzled fur. It is well they have this protective coat, for birth occurs in January or February. Several litters are produced through late spring, when breeding presumably is curtailed.

While the European hare may be an interesting addition to the fauna of our eastern states, the peril of an unwise introduction of this sort must remind us of past disastrous liberations. The European hare does provide fine sport for the hunter, although the hare can quickly outdistance the hounds. Bagging this swift animal is not easy, but the lucky hunter can be assured of a delicious meal.

6

Rodentia

(Rodents or Gnawing Mammals)

The rodents are a cosmopolitan group, found the world over; the exotic house rat and house mouse have been introduced into the United States within historic times by man. For the most part, rodents are terrestrial, fossorial, arboreal, and semiaquatic. Their morphology is generalized, particularly as regards the brain and placentation. Rodents are best characterized by the chisel-like incisors, a single pair to each jaw, and the absence of canines. There is thus a wide space between the incisor teeth and the molars or cheek teeth. The rodents far surpass all other mammalian orders in the number of genera and species, and also in the actual number of individuals. Some, such as our western ground squirrels, introduced rats and mice, pocket gophers, and field mice, often cause great loss to agriculture. On the other hand, the muskrat and beaver are valuable fur bearers, and squirrels provide us with sport and have a very tangible aesthetic value.

Key to the Families of Eastern Rodents

  A. Fur with stiff, spine-like bristles . . . .*Erethizontidae*
AA. Fur soft, without spines.
      B. Tail heavily furred (squirrels) or scaly and flattened horizontally (beaver).
          C. Tail densely furred, feet not webbed . . . .*Sciuridae*
         CC. Tail flat, broad and scaly, hind feet well webbed . . . . . .*Castoridae*
     BB. Tail not well furred nor spatulate.
         D. Hind legs much elongated for leaping, tail at least one-and-a-third times length of head and body, upper incisors with prominent longitudinal groove . . . .*Zapodidae*
       DD. Hind legs and tail not excessively elongated.
           E. A burrower, with prominent fur-lined external cheek pouches . . . .*Geomyidae*

EE. No external cheek pouches. Form rat- or mouselike, tail naked or
furred, but not heavily furred as in squirrels.
　　F. Large aquatic muskratlike rodent, with round tail and four
　　　　cheek teeth (Nutria) . . . .*Capromyidae*
　　FF. Three sets of check teeth.
　　　　G. Molar teeth with tubercles in three longitudinal series
　　　　　　(Figure 6.61) . . . .*Muridae*
　　　　GG. Molar teeth with tubercles in two longitudinal series
　　　　　　or flat-crowned (Figure 6.61) . . . .*Cricetidae*

SCIURIDAE

(Woodchucks
and Squirrels)

Key to the Genera of the Family Sciuridae

A. Incisors white, top of skull flat, tail short and bushy, size large . . . . .*Marmota*
AA. Incisors yellow, top of skull convex, size not large, less than two pounds and
having total length less than 600 mm.
　　B. Lateral furred membrane joining fore and hind limbs, tail depressed
　　　. . . . .*Glaucomys*
　　BB. Lateral membrane absent.
　　　　C. Back usually with prominent stripes or barred color pattern (obscure in
　　　　　Spermophilus franklinii)
　　　　　　D. Dorsum with four pale stripes, five cheek teeth in upper jaw
　　　　　　　. . . . .*Eutamias*
　　　　　　DD. Dorsum with two pale stripes, four cheek teeth in upper
　　　　　　　jaw . . . .*Tamias*
　　　　　　DDD. Dorsum with five or more distinct dark stripes which enclose a
　　　　　　　line of pale spots; or, in *S. franklinii,* with obscurely spotted or
　　　　　　　barred coat pattern . . . .*Spermophilus*
　　　　CC. Back without stripes or barring, usually of one color, the tail long and
　　　　　bushy.
　　　　　　E. Size large, more than 400 mm . . . .*Sciurus*
　　　　　　EE. Size less than 400 mm; reddish above, white below . . . .
　　　　　　　. . . . .*Tamiasciurus*

Eastern Chipmunk. *Tamias striatus* (Linnaeus)

DESCRIPTION.　The eastern chipmunk is a small striped ground squirrel,
with prominent rounded ears and flattened, well-haired tail. The dorsal pattern
is marked by five dark stripes and two white or buffy stripes with rusty or pale
brown rump. It has prominent internal cheek pouches. A specimen from

Marietta, Georgia, is colored thus: general color above dark, the head peppery, dark yellow brown, facial stripes and cheek clay-colored, lips saffron; broader dorsal stripes gray, the individual hairs barred with brown; mid-dorsal stripe black, fading into rump color which is rich chestnut; the two light dorsal stripes cinnamon buff; sides of body saffron; tail fuscous black, peppered with gray, rusty below; hind feet ochraceous buff, the thighs russet. The belly is white, often suffused with ochraceous buff (Figure 6.1).

Average measurements of ten specimens from North Carolina, Georgia and Kentucky are: total length, 232 mm; tail, 84 mm; hind foot, 35 mm. The darkest colored individuals occur in Kentucky and Tennessee and in the Carolina Mountains. Populations containing the smallest and palest individuals are found in the northeast. Ten specimens from Ithaca, New York, average: total length, 233 mm; tail, 87 mm; hind foot, 34 mm. The average weight of twenty adults was 86 (72–100) grams. Howell gives the average measurements of eleven adults from Fort Snelling, Minnesota, as: total length, 268.4 mm; tail, 101.3 mm; hind foot, 36.6 mm. Seven New Jersey adults averaged: total length, 251 mm; tail, 90.5 mm; hind foot, 33.5 mm. The average weight of ten adults was 89 (75–101) grams.

DISTRIBUTION. The eastern chipmunk occurs over most of the eastern United States south to Louisiana, northern Georgia and Alabama, the western Carolinas, and Virginia (Figure 6.2). This species varies in coloration considerably throughout its range, but individuals from any one locality usually show remarkably little variation.

Figure 6.1. Eastern chipmunk, *Tamias striatus*.

Figure 6.2. Distribution of the eastern chipmunk, *Tamias striatus.*

HABITS.   The half-cleared forest and the farm dooryard are equally attractive to the chipmunk. It delights in the open woods, and frequents stone walls, half-rotted logs, and the thick underbrush which covers such places. Here it spends its active daylight hours; it repairs to its subterranean quarters with the approach of dusk.

Wherever chipmunks are at all abundant, their animated ''chuck'' can be heard in every woodlot. The call is taken up by several until the woods resound with their lively chatter. If a Cooper's hawk should strike at one, it flees to the earth or stone pile with a startled whistle, leaving the silent woods to the disappointed raptor.

The chipmunk is an expert climber and often may be seen in small trees. It shows a certain wariness when on slender limbs, and never exhibits the carefree deviltry of the tree squirrels.

The burrows are lengthy and complicated, although there is no evidence of the excavated earth which must be removed in order to make these extensive chambers. It is probable that the dirt is carted off and scattered by the chipmunk. These holes descend abruptly for several inches, then level off, occasionally covering thirty feet or more, the exit and entrance often within a foot of one another. Somewhere within this tortuous burrow a large bulky nest of leaves is made, where the chipmunks not only sleep but bear their young.

During the long Indian summer of the north, these active little ground squirrels are busy garnering stores of nuts, seeds, and other edibles for the dark winter ahead. In the northernmost part of their range, their chucking notes may be heard well into October. Finally a time approaches when the woods are silent, the leaves cover the forest floor and the chipmunks are drowsing away the weeks in their snug retreats. If December or January experience a mild spell, the snowy woods reveal countless tracks of these awakened beasts, and it is then that the labors of the past fall are enjoyed, the storehouse gradually shrinking.

By mid-March the males are all above ground on pleasant days, and by early April they have found their mates. The females carry the three to five young for thirty-one days, and another month elapses before the little ones venture into the green world. Some produce young in late July or August, but it is suspected that these are last year's young which have failed to breed in April.

Small birds, mice, snakes, snails, slugs, insects, and other small animal life are all legitimate prey of the chipmunk, but its chief reliance is on small seeds, berries, fruits, and nuts. In the summer its face may be stained with blackberries, raspberries, shadberries, and other juicy morsels, while in the fall its crammed cheek pouches may hold a dozen or more beechnuts. Often one can hear the grating noise, also made by fox and gray squirrels, caused by the teeth chewing open the hickory nuts. In Indiana, chipmunks are often seen in fall, several to a tree, scampering up and down a tall oak gathering acorns, many of which are heard falling to the ground.

Large snakes, hawks, owls, foxes, bobcats, and the house cat are all foes of the chipmunk, but the weasel is perhaps the most dreaded. Its slim body enables it to pursue the chipmunk into the innermost recesses of its burrow.

Chipmunks occasionally pilfer small bulbs from the garden, but their lively manner and pretty ways add much to the enjoyment of the outdoors.

Least Chipmunk. *Eutamias minimus* (Bachman)

DESCRIPTION.   This small, long-tailed chipmunk has five black stripes on
its back, the middle one reaching from crown to base of tail; four pale stripes
between the black ones, the outer of these being whitish; a whitish stripe
above and below the eye; sides of body orange brown or tawny; the shoulders
often bright rufous; belly grayish white; tail pale brown, the hairs near their
tips marked with black. This little chipmunk differs from *Tamias* in its much
smaller size, the narrower and closer black stripes of the back, and the relatively longer tail, which is usually held upright (Figure 6.3).

Measurements of fifteen adults from northern Wisconsin average: total
length, 202 (183–223) mm; tail, 87 (81–95) mm; hind foot, 31 (28–33) mm;
weight, 40–70 grams.

DISTRIBUTION.   The least chipmunk occurs in north and south-central
Wisconsin and the Upper Peninsula of Michigan (Figure 6.4).

HABITS.   The least chipmunk is universally distributed within its range,
appearing to shun only the thick evergreen forests, and even in such situations
a few may occur. It does not favor the timber so much as its larger cousin,
Tamias, but may be found along the borders of the forest and the shore line of
lakes and it may make its home in pasture land. This species delights in rocky

Figure 6.3.   Least chipmunk, *Eutamias minimus.*

Figure 6.4. Distribution of the least chipmunk, *Eutamias minimus*.

cliffs, open stands of jack pines, and river bluffs, where its numbers may be beyond count.

This little species is a fearless creature, visiting the camp site and making a fascinating nuisance of itself. It will feed from one's hand upon an hour's acquaintance, and religiously explore every package of food, tearing open the sacks of prunes, raisins, and other camp stores.

Eutamias is an expert climber, often ascending trees or bushes to sun itself during the cool of the early morning or actually using a suitable stub or crotch in which to make its nest, after the fashion of the true tree squirrels. Indeed, it has been recorded as rearing its young in an outside nest some feet above the ground.

The young, numbering five to seven are born early in May, and remain with the parent for fully six weeks or longer. They are born in a warm nest constructed of leaves in a burrow among the roots of some windblown pine or other safe subterranean retreat, or actually in a shredded bark nest hidden among the evergreen foliage of a pine or spruce. Gestation is about thirty days.

Great stores of food are cached from midsummer well into the fall. The busy creature, its long tail held erect, garners stores until the bulging cheek pouches appear ready to burst. The food consists of a great variety of items, including nuts, especially the hazelnut, fruits, berries, grasses and green leaves, fungi, snails, insects, and quite probably small birds and lesser mammals that it can catch and overcome.

When disturbed this little animal is both more active and more noisy than the eastern chipmunk. Its most common call is a series of chip notes, but, like its larger cousin, it has a repertoire of calls. In some places in Wisconsin the two species of chipmunks occur together, apparently in harmony.

Although this species does hibernate, the early snows and the icelocked ponds still find it about, weeks after its larger cousin has gone into the winter dormancy. Even in northern Michigan and the Superior region, its tracks may be seen on the winter snows, and mild spells during December and January may bring it from the subnivean depths to scamper about.

Few mammals are so confiding as this little beast, and its large numbers and the ease with which it can be observed give reason to wonder why so little has been recorded about its habits.

Woodchuck. *Marmota monax* (Linnaeus)

DESCRIPTION.   The woodchuck is grizzled brown or grayish brown, the tail is relatively short and well furred, the incisor teeth are *white,* and the ears are short and rounded. It is grayish brown above, tinged with buffy; underfur blackish brown at base, tipped with pale gray or ochraceous buff, more strongly marked on rump; guard hairs broadly tipped with buffy white; top of head and face clove brown to fuscous, the tips of the hairs not lighter; lips, chin, and sides of face buffy white; feet and legs blackish brown to nearly black; tail colored as legs but hairs tipped with buffy white; underparts light, ochraceous buff to buffy white. Measurements of eight individuals from Virginia average: total length, 590 mm; tail, 142 mm; hind foot, 86 mm. In the north, this species is strong reddish above and below, and the species attains its smallest size to the north, with five adults from Maine averaging total length, 505 mm; tail, 118 mm; hind foot, 75.5 mm. In New York, two rather distinct phases occur. One is a light form, the hairs of the dorsum being broadly tipped with light buff. In the dark form, the long hairs are tipped with chestnut brown; the pectoral hairs are mars orange. Melanism is quite common in this species, and albino specimens are by no means rare. Average measurements of 198 New York adults in the flesh are: total length, 555 mm; tail, 125 mm; hind foot, 76 mm; adults weigh 6-10 pounds (2720-4536 g), with an occasional individual weighing 12-14 pounds (5443-6343 g). Average measurements of nine adults from eastern Massachusetts are: total length, 531 mm; tail, 136 mm; hind foot, 77 mm (Figure 6.5).

DISTRIBUTION.   The woodchuck occurs throughout the northern part of our area south to northeastern Mississippi, central Alabama, northeastern Georgia, and southern Virginia (Figure 6.6).

HABITS.   When our country was first settled woodchucks were probably scarce. With the cutting of the forests, and the resultant meadows, open fields and arable lands, it increased greatly, until today it is a familiar sight from nearly any highway in eastern United States. The woodchuck prefers a slightly rolling country, interspersed with ridges, although it is found abun-

Figure 6.5. Young woodchucks, *Marmota monax.* Photo by W. J. Schoonmaker.

dantly in the forested areas and wooded groves of the prairie states. It does not altogether shun its primitive habitat, and woodchucks may occasionally be found within rather dense stands of timber, but these are usually solitary individuals.

Woodchucks are most active in the early morning and late afternoon, but they may be about at all ħours, and if much hunted they sometimes feed at night. While they are more or less solitary beasts, two may occupy a burrow, and several will live peaceably together in a small meadow.

The home is an extensive soil burrow; often the entrace is beneath a stone wall or tree stump. Woodchuck burrows can be distinguished from those of other animals by the invariable fresh mound of earth which lies at the entrace. The burrow may have two or more entrances; it sometimes sinks to a depth of five feet and extends thirty or more feet. One or more of the chambers terminate in a blind sac, where a bulky grass nest is situated. There is often a plunge hole, cleverly dug from within so that no tell-tale mound of earth will reveal it. The woodchuck is an accomplished climber, and may

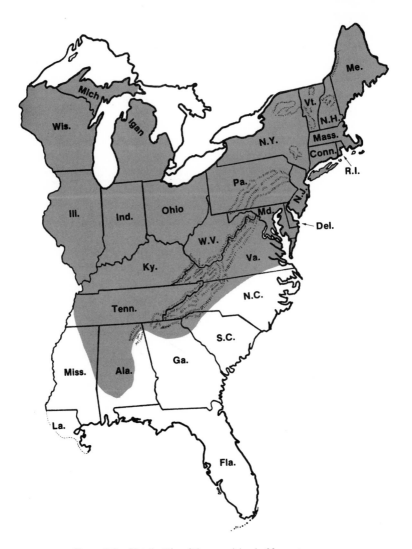

Figure 6.6. Distribution of the woodchuck, *Marmota monax*.

ascend a sizable tree to survey its domain or take refuge when closely pursued by an enemy. It can also swim well.

During the summer the woodchuck stuffs itself with food and becomes extraordinarily fat. As the first frosts blacken tender plants the woodchuck withdraws from the fields and hedgerows and descends into a burrow, often in the woods. While the fall rains give new lusciousness to the green world above, the woodchuck has gone into seclusion, to curl in an inanimate bundle through the long winter. Now the fat stored during the summer is most useful, and is drawn upon during the long period of dormancy. In late February or early March, while snow yet covers the woods, the tracks of the big fellow may be seen wandering from one hole to another, telling of his search for a mate. An occasional individual may sometimes remain active through the winter, at least as far north as central Indiana, feeding on corn remaining on the ground after the harvest.

After a gestation of approximately thirty-one days, the blind, naked young, weighing about an ounce, are born, in a period between the first of April and early May. They usually number four or five, and remain in the den for four weeks, until the eyes open. When a month old, the youngsters commence to feed on grass, and by early July they leave the parent to establish their own shallow burrows (Hamilton 1934).

The principal foods are the various grasses and tender green succulents, as clover, alfalfa, plantain, and various perennials. It delights to follow the rows of sprouting beans, peas, and other early vegetables, and will tear down corn in the milk and eat quantities of apples. Unlike many of the squirrel tribe, the woodchuck seldom appears to eat flesh, although it has been known to pursue poultry and eat insects and snails.

The greatest enemy of the woodchuck is the hunter, who travels the highways and shoots often from a car window, or stalks the fields with binocular and rifle. The farm boy and his dog also take numbers. The deadliest natural enemy is the red fox. Many are caught by Reynard, and toted to the den for the cubs to worry and maul.

The woodchuck spends much time standing upright, ever watchful, near the entrance to its burrow. If suddenly disturbed it will often give a sharp whistle followed by softer chuckling notes as it descends into the burrow entrance, from which it will peek forth.

On "Groundhog Day," February 2, legend says that the groundhog will emerge from its burrow; if it sees its shadow it will reenter hibernation, for six more weeks of winter are at hand. In the northern parts of its range, spring emergence is much later than this, but in central Indiana five woodchuck dens were found to have been recently reopened on February 5, 1971, and shortly after the "Blizzard of 1978," with temperatures hovering near zero, a woodchuck emerged on February 2 or 3, leaving a big mound of fresh dirt on top of the foot of snow still on the ground.

Woodchucks at times cause some damage to the garden, and the hidden burrows and sprawling heaps of dirt and stones which they push to the surface endanger the horse and dull the blades of the mowing machine. Their burrows provide a good home or refuge for many other species, particularly the cottontail rabbit, but also opossums, raccoons, skunks, and foxes, and also various species of mice. The hides are worthless, but the flesh, while coarse, is enjoyed by some.

Key to Ground Squirrels of the Genus Spermophilus

A. Animal with several light stripes, and with spots between the stripes . . . .
. . . .*Spermophilus tridecemlineatus*
AA. No stripes. Squirrel brownish with obscure black flecks . . . .*S. franklinii*

Thirteen-Lined Ground Squirrel. *Spermophilus tridecemlineatus* (Mitchill)

DESCRIPTION.   This species is easily distinguished from any other mammal by the thirteen alternating light and dark stripes on the back. The color above, dark brownish, is interrupted by seven long yellowish-white stripes. The dark area is broken by a central row of rounded spots colored as the light stripes. These stripes are irregularly broken up on the top of the head; face and underparts are tawny brown; the tail is mixed with brown and white, similar to coloration of the back, a central portion black, and the border fringed with buffy white hair tips. Measurements of ten adults from Wisconsin average: total length, 277 (250–310) mm; tail, 99.4 (83–110) mm; hind foot, 38.3 (35–41) mm. Weight is variable, about 110 to 140 grams in June and twice that just prior to hibernation in late September (Figure 6.7).

DISTRIBUTION.   It occurs throughout the states of Wisconsin and Michigan, in northern Illinois and northern Indiana eastward to central Ohio. It was first observed in Lancaster, Ohio, in 1933, and appears to be extending its range in eastern and northeastern Ohio (Figure 6.8).

HABITS.   One of the most conspicuous sights to the tourist driving from western Ohio through Illinois is this scurrying little beast, its short legs carry-

Figure 6.7. Thirteen-lined ground squirrel, *Spermophilus tridecemlineatus.*

Figure 6.8. Distribution of the thirteen-lined ground squirrel, *Spermophilus tridecemlineatus.*

ing it so rapidly across the highway that one fails to see the prominent dorsal markings. Like other mammals which are quick to respond to environmental changes, the striped "gopher" has emigrated beyond its native prairie into the fields and pastures which are the result of deforestation. It is probably much more widely distributed than it was a century ago, for a wider habitat suitable to its needs has been developed by man.

The thirteen-lined spermophile is a sociable little beast, and considerable numbers often occupy a relatively small area on a prairie knoll, in a cemetery, or on a golf course. They rarely venture far from their burrow, sitting up often to make certain that they are not being stalked by some predatory beast. Now and again the tremulous whistle is heard; if danger threatens, it disappears into the nearby burrow. These burrows vary in size with the texture of the soil and the age of the individual. Large squirrels may tunnel to a depth of a foot or more, and the burrow may exceed twenty feet in length, a bulky nest of grasses being prepared in a side chamber. Young animals may make a shallow burrow not more than six feet long, with a single entrance.

Mating occurs soon after the stripers have left their hibernating chambers. The gestation occupies a lunar month; the young are blind, naked, and helpless at birth. It is another four weeks before their eyes are fully open; they are not independent of the mother until about six weeks old. Then the six to twelve young leave the natal chamber, perhaps with regrets, to dig their own home and busy themselves with establishing a thick layer of fat, for winter is approaching and the call to earth is imminent.

By mid-October the striped gopher, resplendent in a new coat, rolling in fat and double its spring weight, is ready for hibernation. Young squirrels may occasionally be seen into late November, and a few are active on warm days throughout the winter. But the stripers belong to that group which best exhibit true hibernation, and usually remain dormant until late March or early April. During this winter sleep a great change occurs. While active, these squirrels may have a temperature from 30°C to 41°C, but in the profound winter torpor their temperature may drop to 2°C. The heartbeat is reduced from a rate of 200 to 350 per minute in active animals to not more than 5 beats per minute in the dormant individual. As much as 40 percent of its weight is lost by a hibernating ground squirrel. The best studies of hibernation in ground squirrels were made by Otis Wade (1930).

Like others of its kin, the striped gopher delights in the sun, and seeks the sanctuary of the earth with nightfall.

These little mammals are rather omnivorous, devouring grains, seeds, succulent plants, berries, insects, birds, small mammals, and even their own kind. A study of the stomachs of 125 thirteen-lined ground squirrels from Indiana indicated clover plants and caterpillars to be two of the most important foods, followed by various kinds of grass and herb seeds, grasshoppers, beetles, and grubs. Only the internal organs are eaten from some of the larger insects. Those who drive slowly will recognize these squirrels feeding on the crushed bodies of their less fortunate kin, scurrying momentarily from the highway as a car approaches. They are known to store quantities of grain in their dens, but when this is utilized is difficult to say. The squirrels may feed on the stored food if they awaken during the winter, but it is most likely used early in the spring.

Many enemies help to keep these little striped squirrels from overpopulating the prairies. Badgers, foxes, weasel, hawks, bull snakes, and other predators are on the alert, and highway mortality is ever a threat to those who choose the border of main traffic lanes as a residence. In late June, 1940, Hamilton counted 107 dead stripers on eleven miles of Iowa highway, and inadvertently ran over 3 in this stretch.

While the thirteen-lined ground squirrel does destroy quantities of grasshoppers, grubs, and wireworms, its detrimental effects cannot be overlooked. This farm pest digs up sprouting corn, sometimes necessitating a second planting. As the wheat and oats ripen, the choicest heads are levelled, a few of the plumpest kernels being eaten. Four or five squirrels to an acre (not a large number) will raise havoc in a ripening field of grain. Occasionally they will enter a garden where they may feed heavily on ripening tomatoes. The species is controlled by means of gas, poisoned grains, traps, and shooting. Whitaker's family collected 147 individuals for hibernation studies in approximately fifteen hours by using water from a hose to force them from their burrows in a thirty-two acre portion of an Indiana cemetery.

Franklin's Ground Squirrel. *Spermophilus franklinii* (Sabine)

DESCRIPTION.   Superficially similar to the gray squirrel, Franklin's ground squirrel has a shorter and less bushy tail, a tawny body, and shorter ears. Upperparts are brownish gray and blackish, the rump often dull yellowish or olive; flanks paler; color pattern so arranged as to give a distinct barring or spotted effect, particularly on hind quarters; head usually plain gray or grayish brown; tail grayish, the margins tipped with paler gray or white; underparts dull yellowish white, grayish white or pinkish buff. Measurements of ten Illinois and Wisconsin specimens average: total length, 380 (351–401) mm; tail, 143 (133–156) mm; hind foot, 52.4 (49–55) mm. Weight in spring is 350–450 grams; in the early fall, prior to hibernation, 500–700 grams (Figure 6.9).

DISTRIBUTION.   East of the Mississippi this species is found in northwestern Indiana, northern and central Illinois, and north into central Wisconsin. In 1867 a captive pair escaped at Tuckerton, New Jersey, and established themselves in the sandy fields. The progeny of this pair were still to be found not far from this point a half century later (Figure 6.10).

HABITS.   Franklin's ground squirrel prefers fields with some hedges or the bushy borders between woods and the open prairie. It is seldom abundant, although it is generally distributed over its range. Its large size and lack of

Figure 6.9.   Franklin's ground squirrel, *Spermophilus franklinii.*

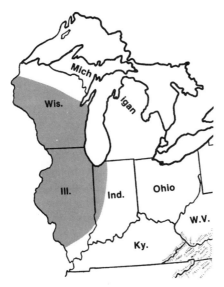

Figure 6.10. Distribution of Franklin's ground squirrel, *Spermophilus franklinii.*

suspicion should make it a favorite object of study; nevertheless, little is known concerning its home life.

The animal may form colonies, although these seldom number more than ten or twelve animals, and considerably fewer may occupy a hedgerow. Their burrows are larger than those of the thirteen-lined ground squirrel, but are usually better concealed, the dirt being scattered from the entrance to leave no evident mound. It is an accomplished climber, and is occasionally seen high up in a tree or brush.

With the approach of cold weather, usually from late September to early November, depending on the latitude, Franklin's ground squirrel, now fat and sleek, retreats to its underground chamber to spend half the year in the deathlike sleep of hibernation. It does not appear above ground until others of its tribe have long since mated. In early April its musical birdlike notes may be heard; possibly this call aids the males in their fevered quest for a mate.

As with all other members of the genus, a single litter is produced, generally by mid-May, after a gestation of twenty-eight days. The young, usually numbering five to eight, are helpless little creatures at birth, and pass nearly a month in the den before appearing above ground. They remain with the parent for several weeks longer, forming a loose-knit community but finally dispersing to seek their own territory as the summer advances. These ground squirrels are peculiar in that they appear in a given region, establish a community, and for no apparent reason desert it, not to be seen again.

The food of this species consists primarily of the prairie grasses and tender herbaceous plants. It causes some destruction to crops, but is not so mischievous as its western cousins. Like all its kinsmen, this species is inordinately fond of flesh, and has been suspected of destroying poultry, eggs, and song birds. It is known to attack mice, the smaller thirteen-striped squirrel, and even small rabbits. Insects of many varieties have been found in the stomachs of *franklinii*. During the summer months, the thickets bordering the prairies provide a plentiful supply of berries, and on these the species may subsist for weeks.

Badgers, foxes, predatory birds, and other enemies levy a toll, but withal this little prairie squirrel maintains itself in some numbers over much of the prairie bordering the lakes. It is a subject worthy of more study, and important new data could doubtless be obtained in the course of a year's planned investigation in an area where it is yet relatively common. It is presently being reduced in many areas.

Key to Squirrels of the Genus Sciurus

A. Tail hairs yellow-tipped; four upper molariform teeth . . . . *Sciurus niger*
AA. Tail hairs silvery-tipped; five upper molariform teeth, the first reduced in size . . . . *S. carolinensis*

Gray Squirrel. *Sciurus carolinensis* Gmelin

DESCRIPTION. This large tree squirrel has a long, somewhat flattened bushy tail; the upper parts of the body are grayish. In southern individuals the upper parts are yellowish brown; the few white hairs of the dorsum giving a slightly grayish cast to the sides of the neck, the shoulders and thighs. The top of the head is colored as the back; the face is clay color to cinnamon buff; ears are yellowish, usually no light tuft at the base; forefeet are gray above; hind feet are raw sienna, the toes gray; the tail is brown at the base, the hairs blackish near the middle and broadly tipped with silvery gray. It is paler and grayer in winter.

Average measurements of ten adults from north Florida and Georgia are: total length, 439 mm; tail, 201 mm; hind foot, 60.5 mm; weight, 400–450 grams (Figure 6.11).

Northern gray squirrels are larger and brighter. Summer pelage is yellowish brown mixed with black on dorsum; cheeks buffy; ears pale yellowish brown; fore and hind feet light brown, the toes gray; sides of body buckthorn brown; tail blackish at base, barred with gray and brown, the tips silvery. Underparts are white, occasionally washed with yellowish brown. Winter pelage is dense and silvery, usually with a pronounced yellowish brown dorsal band which is much paler than in summer; outer ear rich ochraceous orange to nearly pure white; feet pearl gray to grayish white. Melanistic individuals are common, particularly in the northern part of the range. These are quite black above, the hairs near their tips frequently banded with brown; pale brown

Figure 6.11. Gray squirrel, *Sciurus carolinensis.* Photo by Richard Fischer.

below. Average measurements of fourteen adults from central New York are: total length, 487 mm; tail, 235 mm; hind foot, 68 mm; weight 500–710 grams.

Gray squirrels of southern Florida are smaller than elsewhere, and are somewhat lighter and grayer. Individuals from coastal Mississippi and Louisiana are darker than other southern gray squirrels. They are deep yellowish rusty with much black, lending a peppery appearance to the fur; underparts smoky to dark buffy.

DISTRIBUTION.   The gray squirrel occurs in suitable habitat essentially throughout the eastern United States, except for a small area on the west coast of Florida (Figure 6.12).

HABITS.   The bottomlands of the rich Mississippi Valley and the oak forests of the middle Atlantic States are the chosen home of this handsome species, although it is found in lesser numbers much farther north. It was incredibly abundant a century ago, and reports of one man's killing a hundred individuals in a half-day hunt are not unusual. Indeed, the frontiersman had

Figure 6.12. Distribution of the gray squirrel, *Sciurus carolinensis.*

ever to contend with the gray squirrel for his crops, and many a corn patch was rifled of its golden bounty before the harvest season.

Today the squirrel hordes of Civil War days are gone, but the species still thrives in considerable numbers wherever suitable living conditions persist. In the larger parks of great metropolitan areas the squirrels have found a new haven, with abundant food and large nest boxes which serve as shelter.

The gray squirrel is essentially a tree species, venturing to the ground only to collect and bury nuts. It is active throughout the year, braving the northern blizzards to dig in the snow for the multitude of nuts which were buried the previous fall.

A rasping whicker or a harsh squall is given when the animal is disturbed and is a most characteristic note.

Squirrels are most active in the early morning hours and again in the later afternoon, when they forage for food. This trait is well known to hunters, who sit motionless in a favorite spot and wait for the squirrels to become active. Others are on the alert for these handsome creatures. The barred or horned owl, moving on silent wings through the late afternoon, often surprises the unwary squirrel, or an alert bobcat or fox stalks the ground-foraging animal.

A venerable beech or oak is chosen for the home, some natural cavity being enlarged to make roomy quarters for the bulky nest. The squirrels often construct large outdoor nests of leaves, firmly woven together so that they are waterproof. Several squirrels may occupy a single nest. In years when the squirrels are abundant, the numberous nests are a familiar sight in the leafless winter woods.

The great squirrel emigrations of early days were the marvel of the pioneers, and many striking accounts are available. In recent years such emigrations on a somewhat smaller scale have occurred in New England and Kentucky. Gray squirrels in some numbers have emigrated from western Connecticut, even crossing the Hudson River on bridges. The reason for these mass movements has probably been overpopulation in the areas from which squirrels came. There is no evidence that the lack of food has occasioned these rather irregular journeys.

The red and gray squirrel seldom occur in the same area, and it is generally thought that the reds drive away their larger cousins. Some believe that the smaller squirrels emasculate the grays, but there is no direct evidence of this.

During midwinter the gray squirrels mate. The female carries her young about forty days. They are born in the shelter of a cavity or in the large warm leaf nests, and are immature little creatures when they arrive in March. Young squirrels may be observed in late April in the northern part of their range. Another litter is produced in the late summer, especially by older females. In Kentucky a litter is produced in late August and one in mid-February. The young number from one to four, the usual number being two or three.

The distribution of the gray squirrel coincides strikingly with that of the oak and hickory forests. It is evident that the squirrel's mainstay is the mast of these nut trees. In addition, the gray squirrel feeds on a multitude of swelling buds, various fruits and berries, an occasional insect, and perhaps now and then a young bird.

The gray squirrel is a favorite with the hunter, many thousands being killed each year. In most areas, autumn is the hunting season, but Hamilton has seen a young lad in mountainous Kentucky with seven squirrels strapped to his belt in early July. They were shot when repairing to the mulberry trees to feed. In Indiana the hunting season begins in mid-August. It has been claimed that the Revolutionary marksmen first gained their skill with the rifle during their early squirrel-hunting days.

Fox Squirrel. *Sciurus niger* (Linnaeus)

DESCRIPTION.   The fox squirrel is a larger tree squirrel, considerably bigger and heavier than the gray squirrel. The color pattern is exceedingly variable (Figure 1), ranging from black to buffy gray, but the top of head in all color phases is almost invariably black, while the nose and ears are usually light, a creamy white. The fur is coarse, the feet large with naked soles, and the tail flattened and well furred. Three well defined color phases occur, although intergradations between all three are of usual occurrence. The gray phase is characterized dorsally by a buffy gray, the hair tips black; feet and toes are cream to buff, top of head black; nose and ears white; the tail above colored as back, the long hairs black with creamy white tips; underparts yellowish white, ochraceous buff on the underside of the tail, particularly at its base. The buff phase is mixed black and tawny brown above; underparts of tail much brighter and rusty above and below, bordered with orange yellow; underparts ochraceous orange; feet colored as belly; top of head black; nose and ears creamy white. The melanistic phase is black throughout except for the nose, lips, and ears, which are creamy white. Measurements of fourteen Georgia and Florida adults average: total length, 624 mm; tail, 308.5 mm; hind foot, 84.3 mm (Figures 1 and 6.13).

Figure 6.13. Fox squirrel, *Sciurus niger.*

Fox squirrels from southern Florida are similar to typical *niger* as described above but are smaller and much darker both above and below; feet are whiter, not tinged with buff. Nose, lips, and front of face are white; ears are white, with a patch of cinnamon buff at base; head and forelock black, sprinkled with cinnamon, shading on sides to orange cinnamon; hind limbs orange-cinnamon, sprinkled with black; tail above colored like hind limbs, heavily mixed with black and shading on sides to hazel; under surface of tail rich tawny. Harold H. Bailey states that the full black pelage also occurs. An adult male measures: total length, 535 mm; tail, 260 mm; hind foot, 75 mm. It occupies the pine and cypress timbered tracts and the mangrove swamps, from the region of Everglade, Lee County, to the southern part of Dade County. Midwestern fox squirrels are tawny brown grizzled with gray above, somewhat similar to summer pelage of northern gray squirrel, ears bright orange-brown, slightly tufted in winter; cheeks, legs, and underparts pale rufous, or rusty or pale orange-brown; no light nose patch; hairs of tail mixed black and tawny rufous. Individual coloration, however, is variable; some individuals are very dark, almost black, the hairs mixed with gray or tawny. Measurements of twenty-six adult squirrels of both sexes from Ohio average: total length, 535 mm; tail, 244 mm; hind foot, 74 mm; weight, 750–950 grams. Individuals from Louisiana, Mississippi, and Alabama are smaller and of richer ferruginous coloration than typical *S. niger*.

DISTRIBUTION.   The fox squirrel occurs throughout southeastern United States, north through most of Wisconsin and Michigan, western New York, and western and southern Pennsylvania (Figure 6.14).

Fox squirrels are becoming increasingly scarce in the south; collectors should make efforts to secure specimens from the lower Mississippi Valley and the gulf region. The fox squirrel has been almost completely exterminated in the northeastern parts of its range.

The relationships and exact distribution of fox squirrels are not well understood. This group is sorely in need of revision; every effort should be made to save or procure skins and skulls from hunters. The very great color variation in individuals from the same locality make it doubly hard to make generalizations.

HABITS.   The oak and hickory groves of the prairie states are the choice of the northern fox squirrel; in the deep south, where it is not uncommon, open forests of long-leaf pine or open borders of cypress swamps and low thickets are its favorite haunts. This big tree squirrel is seldom found in the deep woods chosen by the gray squirrel, preferring the half-open oak stands, which are often surrounded by prairie and pasture land.

The summer home of the fox squirrel is a loosely built shelter of leaves,

Figure 6.14. Distribution of the fox squirrel, *Sciurus niger*.

conspicuous from a long distance. In the fall, they repair to a cavity in some oak, or make a very substantial structure in a pine, in which the limbs, smaller branches, and a quantity of leaves and bark are firmly molded together. These bulky but well constructed nests are immune to the vicissitudes of the long winter and remain dry and warm during the harshest storms. They may be used for many years, and provide a safe haven against wind, rain, and snow, and also a nursery for the young.

Prenuptial chases occur in this species as early as late December, with first matings occurring through January. A second peak of breeding occurs in May and June. Gestation is about forty-five days. The one to six (usually two to four) young are born primarily in late February and early March, and again in June and July (and infrequently into late August), although pregnant females have been taken in every month of the year. Females born in spring generally produce their young in spring; fall young tend to produce young in fall, although two-year-old squirrels are capable of producing two litters per year. As with other tree squirrels, development of young is slow. The eyes open at the end of the fourth week, and it is seven to eight weeks before they leave the natal chamber, and eleven or twelve weeks before they first venture to the ground and begin to fend for themselves.

The harvest season finds the fox squirrel busily engaged in burying acorns, hickories, and other nuts in shallow graves, where they are recovered in time of need. This is its chief mainstay, and its only food in times of scarcity. But the squirrel does not subsist entirely on mast. The ripening samaras of maples are often eaten, and the fruits of tulip trees are heavily utilized in summer. Berries and fruits in season, the milky golden kernels of the cornfield in late summer, and the bursting buds of spring along with many other items all serve in a measure to supply the needs of this big squirrel. Also, this species will often feed on various fungal foods.

Red Squirrel. *Tamiasciurus hudsonicus* (Erxleben)

DESCRIPTION.   The red squirrel can readily be recognized by its rufous color and small size, about half that of the gray squirrel. In some areas it is referred to as the chickaree, in northern Indiana it is called the piney squirrel and in mountainous West Virginia it is known as fairy diddle. There is a marked seasonal difference in color. In winter a broad rusty red band extends along the entire dorsal pattern, from between the ears nearly to the tail tip; the sides are olive gray, with a sprinkling of black hairs. The prominent reddish or black ear tufts occur only in the winter pelage of the red squirrel; no other eastern squirrel possesses these tufts. The underparts are grayish white, the hairs frequently marked with black to present a vermiculated appearance. The tail above is colored like the back, the outer hairs banded with black and tipped with yellowish rufous; below, yellowish gray, the tips blackish. The summer pelage is duller, more olive than in winter, lacking the bright rufous band. A prominent black line separates the dorsal pattern from the white underparts. Ear tufts are absent (Figure 6.15). Average measurements of twenty-eight adults from western New York are as follows: total length, 310 mm; tail, 120.4 mm; hind foot, 46 mm; weight, 140–220 grams. Red squirrels from Maine are smaller and darker; underparts in winter, gray, much vermicu-

Figure 6.15. Red squirrel, *Tamiasciurus hudsonicus,* winter pelage. Photo by Richard Fischer.

lated with dusky; tail dark colored, with much black, its outer fringe not conspicuously lighter than is the median band. Average measurements of six adults from near Moosehead Lake, Maine are: total length, 290.2 mm; tail, 121.2 mm; hind foot, 44.5 mm.

A dark race occurs in the heavy spruce and fir forests of the higher southern Alleghenian peaks, distinctly darker on head and sides, and the red of dorsal area a deeper shade; underparts in winter more grayish (less clear white) and more or less vermiculated with dusky; the characters show most strongly in winter pelage. Average measurements of thirty-two adults from the Smoky Mountains are as follows: total length, 318 mm; tail, 132 mm; hind foot, 48 mm.

DISTRIBUTION.   The red squirrel occurs from Wisconsin, Michigan, and Maine south through northern Indiana and Ohio and through the Allegheny Mountains (Figure 6.16).

HABITS.   The ubiquitous red squirrel is well known wherever it occurs, from the deep forests of northern Maine and Wisconsin to the mist-shrouded slopes of the Smoky Mountains. Its rolling chatter and mischievous ways have made it a favorite with all naturalists; as a consequence it is one of our best known mammals.

Figure 6.16. Distribution of the red squirrel, *Tamiasciurus hudsonicus.*

Flashing among the spruce tops of dense forests with reckless abandon, or gnawing beneath the eaves of village homes, it exhibits utter disregard for a stereotyped habitat and occupies any area where food and shelter beckon. It spends a great deal of time on the ground, dashing from one tree to another, or digging industriously for stored cones or buried maple seeds. The chickaree actually tunnels into the soil, and will make nests in raised forest knolls which are sufficiently dry. Piles of pine cone remnants are everywhere in occupied coniferous forest.

Red squirrels vary greatly in number. Some seasons the woods appear full of the saucy little fellows; at other times an entire summer may pass with only one or two being seen, and these are strangely quiet.

The home may be a nest of grass and bark in a deserted woodpecker chamber or a natural cavity in some venerable hickory or oak. Nests may also occur in underground burrows, rock piles, and fallen trees, or simply as leaf nests in standing trees. If a grape tangle is at hand, the red squirrels construct their nests of shredded grape bark. These they occupy throughout the year. These chipper little beasts are active except during the most desolate periods of winter, and even at such times they may be about.

In late winter the males may be seen pursuing their mates, chasing and scrambling about the trees and over the ground with great celerity. After a gestation of about thirty-eight days, the three to seven (extremes one to eight) young are born in a relatively immature state. They grow more rapidly than ground squirrels, and when three and a half weeks old have acquired the characteristic markings of the adult. The eyes open on about the twenty-seventh day and the young are weaned at about five weeks. Chasing and mock fighting are common by about the thirty-eighth day, and the food storing habit is manifested at about six to seven weeks. Many young leave the nest before they are fully able to care for themselves and are found in a weak and helpless condition. Some females produce a second litter in August or September. Hamilton has found nest young in late October in New York.

All tree squirrels are lavish in their feeding habits, discarding far more than they eat, and storing an excessive amount. This is an adaptation to their needs, for it teaches the squirrel to acquire a taste for many types of food and to know where all such may be found if any one source is threatened. James N. Layne (1954) found that acorns, hickory nuts, and beechnuts are some of the year-round staple foods of this species, but hemlock cones, tulip tree and sycamore seeds were heavily utilized in some situations. However, various berries, the swelling buds of maple and elm, fruits of sumac, the seeds of many conifers, many species of fungi, an occasional nestling or a clutch of eggs, and many other items are included in its fare. In the early fall the green cones of pine are cut and buried in the damp earth, for if they were allowed to ripen the seeds would quickly be wind-scattered and lost to the squirrel. Even the deadly amanita is eaten by the squirrels, and various fungi are deftly cached in trees where they dry and are available well into the winter. In the spring the chickaree slits the bark of maples to suck and nibble on the sweet icicles thus formed. There can be no doubt that they pilfer the nests of song birds, but the loss is not serious. Red squirrels have been known to kill and partially devour young cottontails.

The quickness of this little pirate does not always save it from enemies. The marten of boreal forests is said to depend on the squirrel for much of its food, and the fisher, the bobcat, the larger hawks and owls, and other predators feed upon it. It cannot overcome the fire terror. In the far north, where fur animals are becoming increasingly scarce, a price has been placed on its hide, and the chickaree is now sacrificed on the altar of fashion.

Key to Flying Squirrels of the Genus Glaucomys
(The two species are difficult to distinguish.)

A. Brownish pelage; total length of animal generally over 260 mm. Skull generally over 36 mm . . . . *Glaucomys sabrinus*

AA. Grayish pelage; total length of animal generally less than 260 mm. Skull less than 36 mm . . . . *G. volans*

Southern Flying Squirrel. *Glaucomys volans* (Linnaeus)

DESCRIPTION.   This flying squirrel is characterized by small size, a prominent loose fold of skin extending from the wrists to the ankles, and a broad, flattened, well furred tail, the sides of which are nearly parallel. The fur is soft, dense, and silky, and the eyes large, black, and lustrous. The hairs above are slaty gray at the base, the tips varying from gray to drab or pinkish cinnamon; sides of body darker; tail above uniform soft gray, lacking the cinnamon tips, below pinkish cinnamon; underparts pure white, occasionally creamy white. Toes are white in winter. Measurements of twelve adults from Ithaca, New York, are: total length, 226 mm; tail, 100 mm; hind foot, 29 mm; weight, 45–70 grams. In the southern states, this species is similar in size but darker. Three adults from the Smoky Mountains average: total length, 237 mm; tail, 102 mm; hind foot, 31 mm; weight, 60–100 grams (Figure 6.17).

DISTRIBUTION.   This species occurs throughout most of the eastern United States except for northern Wisconsin and the northern part of the upper peninsula of Michigan, much of northern New England and New York, and extreme southern Florida (Figure 6.18).

HABITS.   From the oak, hickory, aspen, and maple forests of the north to the moss-draped live oaks and gums of the deep south, the southern flying squirrel occurs. It escapes the notice of man largely through its ghostly habits, for it becomes active only after the shadows have deepened and black night cloaks the woods. This species may often actually outnumber the larger and more conspicous gray squirrel. Often the only intimation of its presence is a luckless victim brought to the doorstep by a cat, or the lifeless but beautiful body taken from a rat trap. One can often frighten one of these little beasts from its retreat by tapping with a stick on dead stumps six to twenty feet high and containing woodpecker holes.

The flying squirrel is a sociable little beast, several sharing a crowded nest during the day. Whether these groups are family clans or just neighbors has not been determined. They emit sundry little notes, some resembling the drowsy twitter of sparrows. These calls may be heard in the forest on a late summer evening and probably serve as a means of communication.

While it cannot fly in the strict sense, the flying squirrel progresses from one tree to another by means of a glide, accomplished by the outstretched lateral membrane. Launching itself from the uppermost parts of a tree, it sails through the air in a descending curve, checking its flight by a gentle upward swing and landing head-up on a neighboring tree. Up this it scrambles, and it glides off through the forest in a series of leaps until the foliage hides its flight. Close observation of captive squirrels indicates that these creatures can turn at right angles from their line of flight during a glide, and can actually elevate or depress the level of the glide by manipulating the membrane and tail.

Figure 6.17. Southern flying squirrel, *Glaucomys volans.* Photo by Richard Fischer.

Winter holds little terror for these densely furred squirrels, although long periods of intense cold make them inactive. At such times several will gather together in a snug ball, to afford one another protection from the low temperatures. Thus ensconced, they remain inactive for several weeks at a time.

Their food includes the usual assortment of berries, nuts, and other delicacies, and they do not disdain insects. Hamilton has caught them in traps set for weasels and baited with a dead mouse. Hickory nuts and the acorns of the white oak are staples, and in the south; pecans and the many berries of woody shrubs must receive their attention. These squirrels victimize the nests of the small birds, and their propensity for flesh makes them potential predators of the passerine species.

In the northern states the breeding season commences in late February or early March. After a gestation of forty days the two to six young are born in an undeveloped condition, quite blind and naked but with a prominent lateral fold of skin which foreshadows the flying patagium. Generic characters may be recognized at one week; when four weeks old the young have opened their eyes and resemble the parents. A second litter is produced in July or August. Farther south (Kentucky) breeding commences in January and the first litter is produced in early March, while the second litter is born during late August.

The larger mammalian predators, owls, and an occasional hawk take toll of these pretty squirrels, but none can prove more relentless, particularly about the surburban towns and villages, than the domestic cat. Almost every cat owner in the rural districts has found the tail of some luckless flying squirrel on the doorstep.

Their gentle and confiding disposition makes these animals ideal pets, although those who would watch them during the hours of their frolic must nap during the day.

Figure 6.18. Distribution of the southern flying squirrel, *Glaucomys volans.*

Northern Flying Squirrel. *Glaucomys sabrinus* (Shaw)

DESCRIPTION.   A handsome species, the northern flying squirrel is larger and more brightly colored than *volans,* which occupies much of its range. It may be distinguished from the smaller *volans* by the browner dorsal pelage, the plumbeous base of the belly hairs (occasionally these hairs are white throughout), and the greater size. In winter, the pelage is colored thus: face grayish white; top of head and back pinkish cinnamon or cinnamon; membranes grayish black, underparts white, often with pinkish or buffy cast, bases of hairs usually dark; tail smoke-brown to hair-brown, light pinkish cinnamon below. The feet are paler than the membrane. The large round ears usually have a small point on their apex, noticeable in fresh specimens. It is much darker above in summer, and the underparts are suffused with buff. Measurements of eight adult New York specimens are: total length, 271 mm; tail, 123 mm; hind foot, 36 mm. The weight varies from 125 to 200 grams. Individuals in the lower Appalachians are darker; tops of fore and hind feet fuscous; cheeks clear gray, not buffy gray; tail darker and more extensively clouded on terminal third than in northern individuals.

DISTRIBUTION.   The northern flying squirrel occurs in the northern states in our area and south in the Appalachians (Figure 6.19).

HABITS.   Dense coniferous forests, stands of venerable yellow birch and hemlock, and less often the more open woods of maple and beech are the haunts of these handsome squirrels. One may see the nut middens left on stumps or about the bases of trees in the broken sunlight of the forest, but only by taking along a powerful flashlight and exercising sufficient patience on the

Figure 6.19.   Distribution of the northern flying squirrel, *Glaucomys sabrinus*.

night watch will the naturalist be rewarded with a sight of these graceful creatures.

If one is to see a flying squirrel in daylight hours, one must pound, or better, scratch, on every stub or hollow limb that can be found. In the retreats made by the pileated and other large woodpeckers, these squirrels build a nest of shredded bark. Either they sleep soundly, or they hesitate to emerge from the comparative safety of the home during daylight hours, for one may weary before the inhabitant of a stub will take alarm.

Winter holds no terrors for the northern flying squirrel. Even during the severest weather, when the mercury tumbles well below zero, they are abroad to rifle the cubbies and spoil the trapper's marten sets. We have been informed by Adirondack trappers that it often becomes necessary to clean out the flying squirrels in the neighborhood before the valuable fur bearers can be taken.

These squirrels glide from tree to tree, volplaning from the higher branches or trunk of one toward the base of another, until, just before terminating the glide, the squirrel deflects its course and sails gently upward, to alight and scramble into the concealment of the foliage above. Bachman credits the flying squirrel with a "sail" of fifty yards, and although we have never witnessed such a tremendous glide, such a distance is not at all improbable.

As night descends, these soft-furred squirrels repair to the ground and forest litter, searching for nuts, fungi, beetles, and other delicacies. We have no thorough account of their dietary habits, but there is no reason to suppose that it varies from that of the red squirrel. Inasmuch as the country they inhabit has long and severe winters, where deep snows may cover the ground for months, the flying squirrel must store quantities of food.

The breeding season commences in late winter and the young are brought forth from late March to early May, the time of birth depending somewhat on the latitude. The gestation period is about forty days. The young are weaned at about forty-five to fifty days. The length of time that the young are dependent upon the parent has not been determined. Another litter is produced in midsummer, the young attaining a good size by mid-September. The nest for the reception of these young is usually in some hollow stub, although outside nests, prepared of shredded bark and leaves, may be built well removed from the ground in the crotch of a conifer. A large bird nest, capped with shredded bark and leaves, may serve as a shelter.

GEOMYIDAE
(Pocket Gophers)

The pocket gophers are short-furred burrowing rodents, characterized by prominent, fur-lined external cheek pouches, small eyes and ears, prominent claws on the front feet and a naked tail. These animals are admirably modified for a subterranean existence. The skull is massive and angular, with slightly protruding incisors. Each upper incisor of *Geomys* has a large groove on its anterior surface and a smaller groove toward the inner side of the tooth.

Pocket gophers are confined to North and Central America, ranging from Saskatchewan to Costa Rica. In the eastern United States the genus *Geomys* is restricted in its distribution, occurring only in the southeastern states and in Indiana, Illinois, and Wisconsin.

Several species and subspecies of pocket gophers have been described from the southeastern United States. We realize that even fairly small rivers can serve as primary isolating mechanisms in pocket gophers, but we do feel that there are more recognized forms than are warranted, and that the previously recognized southeastern species should be considered as subspecies.

Key to the Eastern Pocket Gophers, Genus Geomys

A. Nasals strongly constricted near middle . . . .*Geomys pinetis*
AA. Nasals not constricted near middle . . . .*G. bursarius*

Plains Pocket Gopher. *Geomys bursarius* (Shaw)

DESCRIPTION.   This is a large pocket gopher with a medium-to-long, scant-haired tail, the tip of which is nearly naked. The coloration is dark liver brown or chestnut above and below, somewhat paler on the belly; the coloration closely paralleling the soil color of the particular locality. The forefeet are white with the hind feet soiled white; the hairs of the tail are usually brown on the basal half and white on terminal half.

This is a large pocket gopher. Some males measure more than 300 mm in total length and weigh 500 grams. Average measurements of five adults from nothern Wisconsin are: total length, 288 mm; tail, 86 mm; hind foot, 36 mm. Individuals from Illinois and Indiana are distinctly slate gray with no trace of brown or brownish on back; hairs tinged with light brown or white on belly. They are large; total length of males, 265–322 mm; females, 253–273 mm.

DISTRIBUTION.   East of the Mississippi *G. bursarius* occurs in west-central Wisconsin, central Illinois and northwestern Indiana (Figure 6.20).

HABITS.   Areas of brown sandy loam or prairie loam sparsely shaded by trees and the open pasture land in agricultural areas are the homes of these large pocket gophers. Intensive cultivation and new roads apparently disturb existing colonies and possibly limit range extension of the species.

Pocket gophers are solitary little beasts, living an unsocial and indeed an antisocial existence in their underground chambers. When two are placed

Figure 6.20. Distribution of the plains pocket gopher, *Geomys bursarius.*

together they frequently engage in savage combat, which often causes multiple fractures of the limbs and reduces the body to a pulpy mass of flesh. Even when one meets death in such a battle, examination proves that the incisors, however sharp they may be, have seldom pierced the hide.

Geomys is apparently more active during the summer months, shuttling back and forth in its subway with edibles for the storage chambers. Usually it does not burrow much deeper than a foot or so at this season, but when the winter approaches it is thought to penetrate the soil deeper, digging below the frost and becoming less active. Melting snow often exposes the long earthen cores of solidly packed loam, which is pushed into the snow when the burrow is being freed of dirt. It is probable that only a single litter is produced each season; their increase is thus not spectacular like that of the smaller rodents. Inasmuch as pocket gophers have few natural enemies, even this slow reproductive rate allows them to increase to a point where they are occasionally a serious agricultural pest.

The food is largely if not entirely vegetable. Most important are fleshy roots, bulbs, and the tender green plants, many of which are probably cut off at the roots and pulled into the burrow. Perhaps some plants are also reached from the mouth of a burrow. The roots of small shrubs and trees are not spared. Quantities of such food are stored in the underground chambers against the time when growing plants are difficult to obtain. The pouches are used to transport food, not dirt.

Pocket gophers often become serious pests, particularly when they are attracted to the garden, vegetable patch, or alfalfa field. They are not difficult to control. Special gopher traps have been designed and are very efficient, but if small steel traps of any description are placed in the main burrow, and the hole is covered with cardboard and dirt, to exclude light, the mammalogist can count on securing a series of specimens.

Other than for the gopher snake, *Pituophis,* which enters its burrows, the pocket gopher has relatively few enemies, although a weasel, house cat, or owl occasionally manages to capture one outside its burrow.

The degree of isolation from other species is exhibited by the number of host-specific ectoparasites possessed by these creatures. In Indiana, all six of their major parasites, a flea, a louse, and four species of mites, are host-specific forms.

The obvious mounds are the mark of the gopher. The pushing up of the mounds is the gopher's means of disposing of excess dirt as he extends his burrow system. Often the mounds are in a line, with progressively fresher dirt indicating the newer burrow. The nearly naked tail apparently serves as a tactile organ as the animal moves backward in its burrow. The gopher uses its powerful incisors to cut away the dirt in burrowing; the front feet are used both in digging and, like the hind feet, in moving the dirt backward under the body. When dirt has accumulated behind, the gopher will turn and plow the dirt backward with his legs and head in the manner of a bulldozer. Food is put into the pouch by the front feet through a rapid sweeping motion.

In areas where pocket gophers are numerous, sizable tracts may be found which are quite barren of these animals. These have been termed resting grounds: they probably permit new growth of food, reconsolidation of soil, and possibly disinfection and reduction of excreta.

As with other members of the Geomyidae, water appears to be a formidable barrier to their dispersal, and rivers often mark the natural boundaries of the ranges of the different species or subspecies. Unlike most other small mammals, pocket gophers cannot swim and make only ineffectual movements when tossed into the water.

For an animal which is so widely distributed and known by so many naturalists, it is surprising to note how little is known concerning the reproductive habits of these animals. Pairs are found together only in the spring; after mating it is thought that the male leaves the female to pursue a solitary existence. Two to six (usually four) young are born in the spring, but the gestation period is not known. At birth the young are undeveloped little creatures, but they remain with the parent only until weaned, when they separate to live their solitary lives.

Southeastern Pocket Gopher. *Geomys pinetis* Rafinesque

DESCRIPTION. *Geomys pinetis* is smaller and paler than *G. bursarius.* It is cinnamon brown above, tinged with fulvous and buffy below; the feet and tail are pale buff to white.

Average measurements of nineteen adults (ten males and nine females) from near Augusta, Georgia, are: total length, 259 mm; tail, 86 mm; hind foot, 35 mm. The males are larger and heavier than the females (Figure 6.21).

DISTRIBUTION.   *Geomys pinetis* occurs in the southeastern United States, from Georgia to Central Florida, westward to west-central Alabama (Figure 6.22).

HABITS.   As one travels over the flat highways of coastal Georgia and much of Florida by car or train, irregular mounds of light sand, contrasting sharply with the burnt brush or brown earth of the pine woods, are seen even by the most casual observer. These are the mounds thrown from the tunnels of ''salamanders,'' or the southern pocket gophers. They are at times incredibly abundant, several hundred piles of pale sand dotting an acre of pasture or scrubby field. Wherever the soil is loose and friable and sufficiently high above the water table, *Geomys* sooner or later appears to tunnel his dark way through the yielding earth.

Few mammals are so peculiarly modified for a special habitat. The strong, heavy, and protruding incisors tear at refractory roots or obstacles as the heavy long claws of the forefeet burrow through the ground. When a pile of dirt has choked the burrow behind, the gopher turns and with head and forefeet pushes the load, expelling it through a lateral chamber to the surface. As the pile accumulates, a sizable mound is formed. Eventually another lateral is constructed and yet another pile formed, until the industry of a solitary individual gives the appearance of many at work (Figure 6.23). These piles often mark the course of the tunnel, but when a number have accumulated, it is difficult to follow the direction taken by the subterranean passage. The tunnels are made at variable depths, often only a few inches below the

Figure 6.21. Southeastern pocket gopher, *Geomys pinetis.* Photo by Francis Harper.

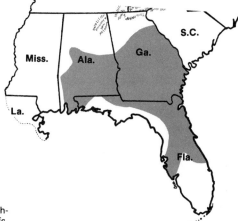

Figure 6.22. Distribution of the southeastern pocket gopher, *Geomys pinetis.*

soil and just as frequently a foot or even two feet beneath the surface. The burrows are often of some size, more than large enough to accommodate their owner. The gopher can run backward with as much agility as forward; the tactile tail tip undoubtedly serves as an organ of touch when the gopher runs backward.

The prominent external check pouches are used only to carry food to the storage chambers, and not to transport dirt, as is often erroneously believed.

Figure 6.23.   Mounds of pocket gophers at Waycross, Georgia, March 25, 1941.

They are filled with incredible rapidity by placing food in them with the fore paws, "a sort of wiping motion which forces it into the open end of the mouth." To empty the pouches the animal presses his forefeet against the sides of the head and brings them forward simultaneously; the contents of the pouches are extruded in a pile in front of the animal.

The breeding habits are not well known, but probably differ little from those of other members of the family. Most species have but a single litter of three or four young; a higher reproductive rate is hardly necessary in the secluded chambers, for few enemies reach them in such a retreat.

The food consists chiefly of roots, fleshy rhizomes, and the green succulents which can be reached from a tunnel opening. Probably no herb is scorned. Considerable food is stored in the underground chambers, far more, it would seem, than the animal could possibly eat. It appears strange that such quantities should be stored, for the gophers are active at all seasons.

In Georgia and Florida the pocket gopher is a pest of no mean consequence in agricultural regions. It is inordinately fond of sweet potatoes, and in addition to destroying quantities of these it attacks peanuts, sugar cane, and peas. In Alabama, however, A. H. Howell (1921) states that the gopher shuns agricultural lands and is of no economic consequence.

CASTORIDAE

(Beavers)

Beavers are large, heavily built animals with broad horizontally flattened tail. They are the largest of North American rodents, large specimens attaining a weight of seventy pounds. The hind feet are prominently webbed, the second toe of the hind foot having a double or cleft claw, which is said to act as an efficient comb in removing ectoparasites. The skull is massive and broad, lacking a postorbital process.

In the New World, beavers are widespread, occurring in suitable localities from Texas to northern Canada.

Beaver. *Castor canadensis* (Kuhl)

DESCRIPTION. The beaver is a large heavy rodent with flattened paddlelike scaly tail and broadly webbed hind feet, the second toe with split claw. Color is uniform dark brown, the dorsal hairs chestnut-tipped, underfur without the bright brown tinge; ears dark blackish brown. Five adult specimens from Maine and New York average: total length, 1170 mm; tail, 412 mm; hind foot, 175 mm. Adults (three years or more) weigh from 30 to 60 pounds (13.6–27 kg), occasionally more (Figure 6.24).

DISTRIBUTION.   The beaver was formerly widely distributed throughout eastern United States but nearly exterminated by 1900. Since then reintroductions and protection have afforded the beaver an opportunity to increase and it is now generally distributed in the eastern states and becoming increasingly abundant (Figure 6.25).

HABITS.   The beaver once occupied much of forested North America wherever suitable watercourses permitted the construction of dams and lodges. Today it is found in regions where it had been absent for more than eighty years.

The engineering feats of the beaver are well known and have been observed by many. The dams, lodges, and less known canals are all remarkable accomplishments. The haunts selected by the beaver for such construction are the wilder lakes and ponds or the gently sloping streams where their dams will create a pond so deep as to permit the accumulation of a substantial food pile for winter use and not to freeze to the bottom. The construction is so devised that branches are placed with the flow of the stream, their butts facing upstream. When a sufficient number are in place, they are secured with mud, sod and even sizable stones. More branches are laid on top of these and the process repeated until the desired pond depth is secured. The lodge is formed by building from the pond bottom, a small island or a gently sloping bank, and is constructed primarily of large branches or the trunks of saplings. A roomy

Figure 6.24.   Beaver, *Castor canadensis.*

Figure 6.25. Distribution of the beaver, *Castor canadensis.*

interior, with a dry platform, provides sleeping quarters and a table where the occupants may repair to feed on the bark of limbs dragged from their food pile. Canals, often of considerable length, are cut to facilitate the transport of branches to the lodge or food pile.

At times beavers make no lodge, preferring a tunnel in the bank. Such homesites are more likely to be used about a natural pond or lake than on a pond created by a beaver dam, and beavers living along rivers routinely use burrows.

Autumn is an active season for the beaver. It is then that they are repairing the dams and lodges and accumulating the food piles that are essential for the winter, when zero temperatures lock them from the shore and growing food. Usually they carry on their activities at night, but at this busy season they work through much of the day if undisturbed.

One or two beavers engage in cutting a tree. When the tree is felled (many are not completely cut through) the branches within reach, up to five inches, or even more in diameter, are cut off, reduced to a length convenient for movement, and transported to the food pile. Often the tree felled catches in the branches of another, and leans far out of reach of the exasperated beavers; and of course every tree which is successfully cut through, if of any size, always retains many branches which the beaver cannot reach. If the bark is not eaten or moved into the water in a week or two, it becomes dry and unpalatable and is not touched by the beavers.

Beavers mate during midwinter, probably in late January or February. The gestation period is suspected to be about three months. The one to six (usually three to five) young are born in April or May, and are remarkably well developed at birth, with well furred bodies and open eyes. They remain in the lodge for a month; at the end of that time they leave to swim and take solid food.

Each colony may consist of a pair of adults and their young of the present and preceding years. Just before the birth of another litter, the two-year-olds are driven out, or leave the colony of their own accord, to establish a new residence, often miles removed. This is the situation as observed in Michigan by G. W. Bradt (1938).

From fall to spring the beaver feeds on bark, and in the north aspen is its staple. Entire stands of ''popple'' may be removed before serious inroads are made on other trees, although maples, willows, alders, apple, birch, and many others are eaten with relish. Conifers are seldom touched. In Louisiana, loblolly pine, sweetgum, silverbell, sweetbay, and ironwood are some of the tree species heavily used. In the summer the aquatic succulents, as pond lilies, bur reed, duckweed, pondweeds, algae, and the fleshy rootstocks of many others are eaten, and the beavers journey to neighboring fields to eat clover, alfalfa, and various herbaceous plants.

Its large size saves the beaver from many carnivores that would otherwise be its enemies. Man, with his traps, is the worst predator. At the present time considerable numbers are taken in New York and Pennsylvania. Large pelts taken there during March 1942, averaged $32, while in 1946 the price had risen to $65. Between 150 and 250 Indiana pelts per year were taken between 1969 and 1971, averaging between $5 and $7, whereas 8,000 New York pelts averaged only $20 in 1976. Suitable protection has resulted in a surplus in many states so that trapping no longer threatens the extinction of this big fur bearer.

CRICETIDAE
(Native Rats and Mice)

The Cricetidae is the largest family of mammals. Members of this family range in size from the tiny harvest mouse to the muskrat. They have no premolars, and never more than three molars on a side. The molars have either flat crowns or tubercles; if the latter the tubercles are always in a longitudinal biserial arrangement and never develop a functional third series on the lingual side of the crown.

Among this large and dissimilar group we find many species with diverse food habits, but most are essentially omnivorous, although they are for the most part adapted to a vegetable diet. Many are of considerable economic importance, being destructive to agricultural crops. The muskrat is the most important native fur bearer.

Key to the Genera of the Family Cricetidae

A. Crowns of molar teeth with two rows of tubercles (Figure 6.61), or with transverse lophs, tail long (at least one-third of total length).
  B. Front of incisors with longitudinal groove . . . .*Reithrodontomys*
  BB. No groove.
     C. Molariform teeth with transverse lophs . . . .*Sigmodon*
     CC. Molariform teeth with two rows of cusps.
        D. Size small, seldom exceeding 200 mm; tail bicolor.
           E. Ears and body bright golden or ochraceous; posterior palatine foramina nearer to interpterygoid fossa than to anterior palantine foramina . . . .*Ochrotomys*
           EE. Ears dusky, contrasting slightly with body color. Posterior palatine foramina midway between openings mentioned above . . . .*Peromyscus*
        DD. Size large, usually more than 200 mm, tail not bicolor . . . .
           . . . .*Oryzomys*
AA. Crowns of molar teeth with loops or irregular triangles, these being particularly well marked in the upper jaw.
  F. Tail long (nearly half the total length) and usually well haired, belly white or creamy, size more than 250 mm; ears large and prominent . .*Neotoma*
  FF. Tail scaly or short and haired, and not bicolor; belly not white, ears usually hidden in fur.
     G. Tail long, flattened vertically, size more than 300 mm . . . .*Ondatra*
     GG. Tail round or terete.
        H. Size large, usually more than 250 mm; tail relatively long, at least one-third total length, similar to small muskrat in appearance . . . .*Neofiber*
        HH. Size smaller, usually less than 200 mm; tail relatively short, less than one-third total length.
           I. Incisors with faint shallow groove on their outer anterior surface (Figure 6.64), tail very short . . . .*Synaptomys*

II. Incisors not grooved.

J. Palate ending in shelf; fur reddish above, contrasted with grayish brown sides . . . .*Clethrionomys*

JJ. Palate not ending in shelf; not reddish . . . . .*Microtus*

Marsh Rice Rat. *Oryzomys palustris* (Harlan)

DESCRIPTION.   The marsh rice rat has a ratlike form; it is superficially much like the Norway rat, from which it can be distinguished by the two rows of tubercles on the molars. It is gray or brown above, much paler below; the tail is long, slender, and scaly, sparsely haired. The body is grayish above, mixed with black and some brown hairs, slightly darker on the mid-dorsal area; face paler; buffy or with yellowish brown wash in Georgia specimens; underparts grayish white, the fur soft and rather woolly; feet white above; tail dark brown or blackish above, slightly paler below, but not bicolor. Average measurements of seven adults from Chincoteague, Virginia, are: total length: 252 mm; tail, 121 mm; hind foot, 30 mm. A large male from the Okefinokee Swamp of Georgia measures 282 mm, 135 mm, and 32 mm (Figure 6.26). Weights average 45–80 grams.

Individuals from peninsular Florida are larger and more tawny. They are grizzled grayish brown to tawny olive, richest and darkest on the rump; underparts white, sometimes suffused with buff. Measurements of ten adults are: total length, 276.7 mm; tail, 143 mm; hind foot, 34.3 mm. Rice rats from Louisiana and western Mississippi are somewhat pale and have a narrower skull than elsewhere.

DISTRIBUTION.   The marsh rice rat occurs from southern New Jersey (Salem) south through Florida, west to Mississippi and north to southern Illinois, and southern Kentucky (Figure 6.27).

HABITS.   The rice rat occupies a variety of habitats, from the salt marshes of the Atlantic and Gulf Coasts to the clearings of wooded areas well into the slopes of the Alleghenies. It is usually found at any place in its range where there is a sufficient ground cover of grasses and sedges to give it protection from the multitude of enemies which seek to destroy it. Rice rats are amphibious creatures by choice, exhibiting a preference for wet meadows and marshy areas. The same subspecies often shows a very wide choice of habitat, from salt marsh to fairly dense forest.

In the salt meadows rice rats occasionally make extensive and well defined runways, or they tunnel in the banks for short distances. Oftentimes a nest of sedges and grasses is constructed in a low bush many rods from land, but this is a matter of individual choice, for the neighboring shore may support large numbers.

Figure 6.26. Marsh rice rat, *Oryzomys palustris.*

Rice rats are accomplished swimmers. When alarmed, they dive and swim under the water for a great distance. Their presence is evident by the mats of cut vegetation floating at irregular intervals in the tidal waters. Because of their nocturnal habits, rice rats are seldom seen, yet they may be the most abundant of all small mammals in certain parts of their range.

A nest of grasses and weeds may be placed under a mass of tangled debris, or woven into the rushes and aquatic emergents a foot or more above the high water level.

The breeding habits are not fully known, the best account being of Louisiana rats. There they breed from February to November. The one to five young are born after a gestation period of twenty-five days. They are blind and naked at birth, but grow rapidly, the eyes opening on the sixth day. They can run about at this very early age and they are weaned when eleven days old. In the Carolinas the young are born throughout the summer. Specimens taken on Cape Charles, Virginia, during early September were of all ages, from very small individuals to subadults. Several breeding females contained three to five embryos apiece.

Bachman, who described this mouse, named it with reference to its habit of feeding on rice. During colonial times, considerable loss was caused in the rice plantations by *Oryzomys,* which scratched up the rice when newly planted and before it had been flooded. The rats likewise feast extensively upon the rice in its milky state, and continue to glean the scattered grains through the fall and winter. Its staple foods are seeds and the succulent parts of available plants, although H. F. Sharp, Jr., found much animal food, particularly insects and small crabs, to be very important to this species in summer and fall in the Georgia coastal salt marshes. Rice rats are also known

Figure 6.27.   Distribution of the marsh rice rat, *Oryzomys palustris.*

to feed on eggs and young of the long-billed marsh wren. They have also been known to feed on Gama grass and marsh grass (*Spartina glabra*).

The barn and barred owls, marsh hawks and other raptors, mink, weasels, foxes, skunks, and other four-footed predators relish these rats. Much of their habitat coincides with the range of the cottonmouth moccasin, and this big pit viper is said to destroy many rice rats. Water snakes also take their toll.

The rice rat is such a widespread species and often so abundant within its distributional limits that we would expect it to be much better known.

Cudjoe Key Rice Rat. *Oryzomys argentatus* Spitzer and Lazell

DESCRIPTION.   This rat is distinct from all the other members of the subgenus *Oryzomys* in having no tufts of digital bristles projecting beyond ends of median claws on the hind foot; large and widely open spheno-palatine vacuities; the slender skull has long narrow nasal bones, their width is contained in length from 4.6 to 4.9 times. Pelage is silver gray laterally. This is the most recently described eastern mammal. It was named by Spitzer and Lazell (1978) in November 1978.

DISTRIBUTION.   This rat is apparently confined to Cudjoe Key in the lower Florida Keys. The two specimens were collected along the edges of a small fresh water marsh bordered by white mangrove with an understory of *Cladium* in the shallows. Where the water deepened to six inches, the saw-grass was replaced by cattails.

This newly described species is presumably quite rare. It would be prudent for the mammalogist to refrain from collecting in this area, lest the population become endangered.

Key to Harvest Mice of the Genus Reithrodontomys
(Species are difficult to identify.)

A. Tail usually more than 110 percent of body length. Last lower molar with dentine
   in the form of an "S" ....*Reithrodontomys fulvescens*
AA. Tail usually less than 110 percent of body length. Last lower molar with dentine
   in the form of a "C."
   B. A distinct labial shelf or ridge, often with distinct cusplets on first and
      second lower molars ....*R. humulis*
   BB. Without such a ridge ....*R. megalotis*

Eastern Harvest Mouse. *Reithrodontomys humulis*
(Audubon and Bachman)

DESCRIPTION.   In general appearance the harvest mouse resembles a small brown house mouse, with a deep longitudinal groove near the middle of the upper incisors. Harvest mice are the only *long-tailed cricetid* rodents of eastern United States *with grooved incisors*. General color of the upperparts is dark brown mixed with cinnamon, darkest along mid-dorsum; the underparts ashy, tinged with cinnamon; tail not sharply bicolored, fuscous above and paler below; ears fuscous or blackish; feet pale gray (Figure 6.28).

Measurements of fourteen adults from Virginia, South Carolina and Georgia are: total length, 121 mm; tail, 56.5 mm; hind foot, 16 mm; weight, 10–15 grams.

DISTRIBUTION.   The eastern harvest mouse occurs from Virginia, southern Ohio, and Kentucky south. It should be looked for in southern Indiana and Illinois (Figure 6.29).

HABITS.   Waste fields of matted grass and broom sedge, tangled patches of brier, roadside ditches, brackish meadows, and wet bottomlands are the

Figure 6.28. Eastern harvest mouse, *Reithrodontomys humilis;* a captive specimen taken near Folkston, Georgia. Photo by Francis Harper.

haunts of this dainty little creature. It is probably more numerous than the numbers taken by collectors would suggest, but it seldom reaches the abundance of other small rodents. Probably inexperienced collectors have occasionally thrown away this species, mistaking it for a buff-colored house mouse. Any specimen of the latter with buffy underparts which may be caught in southern or midwestern fields and ditches should be critically examined to determine if it might be a *Reithrodontomys*. The presence of grooved incisors will quickly establish its identity.

Harvest mice construct soft little nests of shredded grass and plant fibers; these are placed in the tangled herbage under a protecting fence or at the base of a clump of saplings. Quite frequently the home is made in a small shrub several inches above the ground, or it may be woven in the matted aquatic uprights and rushes above the wet muck. It is usual for the nest to have a single entrance. These nests serve the mice throughout the year, for they do not hibernate. Definite runways are not made by these mice, but they frequently utilize the passageways of the cotton rat.

Little is known of the home life of *Reithrodontomys*. The young, numbering from two to seven (the usual number appears to be four), are able to leave the nest chamber when but ten days old, but probably stay with the mother for a longer period. They commence to breed in March, and pregnant females have been taken in South Carolina as late as December. Gestation is presumed to be twenty-one to twenty-two days. The newborn weigh about 1.2 grams.

Harvest mice feed on the multitude of little seeds and choice green sprouts that are everywhere abundant. Limited storage is practiced; at least the mice garner small stores of grass seeds, such as brown sedge, crab grass, and witch grass.

Harvest mice seldom are sufficiently abundant in cultivated areas to cause concern. They provide a source of food for the larger carnivores and, on the whole, they may be considered beneficial.

Figure 6.29. Distribution of the eastern harvest mouse, *Reithrodon-tomys humilis,* and the western harvest mouse, *Reithrodontomys megalotis* (lighter tint).

Western Harvest Mouse. *Reithrodontomys megalotis*
(Baird)

DESCRIPTION.   Like other members of the genus, the western harvest mouse is recognized by the long slender tail and grooved incisors. The middle of the back is brown, with numerous black-tipped hairs; sides grayish buff;

underparts white; tail bicolor, dark brown above, white below. Measurements of five individuals from LaCrosse, Wisconsin, average: total length, 139 mm; tail, 67 mm; hind foot, 16.8 mm. Over eighty adults from Indiana averaged 126.8 (114–146) mm in total length, 58.3 (50–69) mm in tail length, 16.3 (15–18) mm in hind foot, and 10.8 (9.1–21.9) grams in weight.

DISTRIBUTION.   East of the Mississippi this mouse was known through the 1950s only from LaCrosse and Racine, Wisconsin. The species was first taken in Illinois in 1953, in the extreme northwestern corner of the state. There was a series of reports indicating the species was moving eastward across northern Illinois. The first individuals from Indiana were taken in 1969, and the species is now known to occur in several northwestern Indiana counties (Figure 6.29).

HABITS.   The habits of this species are presumably very similar to those of the eastern harvest mouse. It is a mouse of the open fields, doing particularly well in early stage grassy or weedy fields. Fields of giant foxtail, *Setaria,* are particularly often inhabited, but in Indiana, a huge population occurred in a Newton County rye field allowed to go fallow as wildlife habitat.

Breeding occurs from March through October, with two to six young per litter, but there may be a cessation of breeding during the dry part of the summer.

The more important foods of harvest mice in the rye field in Indiana were rye seeds (over half the diet), followed by moth larvae (about 22 percent). The most important items in other fields were moth larvae (31.1 percent of the volume of food), followed by grass seeds (29.2 percent), especially foxtail grass.

Fulvous Harvest Mouse. *Reithrodontomys fulvescens*
 aurantius (Allen)

DESCRIPTION.   This species can be distinguished from the other eastern harvest mice by its longer tail, which measures at least 80 mm in adults and amounts to over 110 percent of the length of the body. The animal is golden brown, often with a blackish band down the dorsum. The sides of the face are a rich "vinaceous-tawny or ochraceous-orange. The underparts are grayish white, strongly washed with light pinkish cinnamon." From Louisiana, 133 individuals averaged as follows: total length 158 (142–200) mm, tail 88 (80–100) mm, and hind foot 19 (15–22) mm. The weight of 49 individuals averaged 12.1 (8.5–17.8) grams. The above data and quotations are from G. H. Lowery (1974).

DISTRIBUTION. The fulvous harvest mouse occurs, in the eastern United States, only in southwestern Mississippi and in Louisiana (Figure 6.31).

HABITS. This species is very common in Louisiana, where it occurs in old fields and thickets, with cotton rats and least shrews in dry areas, and with the rice rat in damp areas. Nests are built one to three feet above the ground and are of shredded grassy plants in a compact mass about the size of a baseball. Overall, the life history of this species appears similar to that of the other harvest mice. The species feeds mainly on weed seeds, but green vegetation is also eaten. Considering the relative abundance in Louisiana, it is surprising that more information is not available concerning this species.

Key to Mice of the Genus Peromyscus

A. Five plantar pads (pads on soles of feet) (Figure 6.37) . . . .
. . . .*Peromyscus floridanus*
AA. Six plantar pads.
   B. Small size; hind feet usually 18 mm or smaller.
      C. Very light color; on sandy areas of southeast . . . .*P. polionotus*
      CC. Dark grayish; in open areas including cultivated fields of north . . . .
      . . . .**P. maniculatus bairdii*
   BB. Larger; hind feet usually 19 mm or larger.
      D. Tail usually slightly more than half the total length; color grayish with dorsal stripe indistinct or absent . . . .
      . . . .*P. maniculatus* (excluding *bairdii*)
      DD. Tail usually slightly less than half the total length; color brownish, usually with well developed dorsal stripe.
         E. Hind foot usually more than 22 mm; southern species usually inhabiting lowlands . . . .*P. gossypinus*
         EE. Hind foot usually less than 22 mm . . . .*P. leucopus*

*This subspecies is entered in the key because it behaves as a species where it occurs with other *P. maniculatus*.

Deer Mouse. *Peromyscus maniculatus* Wagner

A very interesting but difficult taxonomic problem exists with this species. Two morphological and behavioral types occur, one a long-tailed, large, big-eared woodland form, and the other a much smaller, short-tailed, small-eared field form. The two occur together geographically (although in different habitat) in large areas of Michigan, New York, and Pennsylvania, but apparently intergrade through a series of populations in Wisconsin. Thus, the overlapping populations have well-developed secondary isolating mechanisms and are acting as good species, whereas the Wisconsin populations are not even separated by a primary isolating mechanism. This situation is called circular overlap, and defies logical taxonomic placement. If the two forms are called separate species, then one is in the position of having intergradation between two species in Wisconsin. If they are called separate subspecies of one species, then one has two subspecies occurring together and not breeding. The latter is generally accepted as the more conservative approach and is followed here.

DESCRIPTION.   In Canada, this form attains medium size; the tail is nearly half the total length, with a prominent pencil of hairs at its tip. On an adult in summer, the sides and lateral portion of back are dark brown, the midstripe darker, base of whiskers and orbital region blackish, ears dusky, the edges paler; underparts and feet white; tail bicolor, upperpart dark brown to brownish black, white below. Immature mice are grayish above, the middorsal streak nearly black; tail blackish above; ears white inside, blackish outside, the edges sharply marked with white. Average measurements of fourteen adults of both sexes from Canada are: total length, 183 mm; tail, 87 mm; hind foot, 21 mm.

To the south, in New York and western New England, northern Wisconsin and northern Michigan, deer mice are larger, with longer tails. Color above is brownish gray, the middorsal stripe slightly darker than the rest. Cheeks, side of neck, and lateral stripe are slightly browner than the rest of the body; feet white; tail sharply bicolor, dark gray above, white below, with prominent pencil. Color is sometimes bright brown, approaching *P. leucopus* in character, but generally paler in winter. Average measurements of twenty adult specimens from northern New York are: total length, 191 (175–215) mm; tail, 92.7 (79–106) mm; hind foot, 21 (19–23) mm; weight, 16–29 grams (Figure 6.30).

Eighteen adults from Lynch, Harlan County, Kentucky, taken at an elevation of 4000 feet, average: total length, 178 (164–198) mm; tail, 89 (77–97) mm; hind foot, 20.3 (19–22) mm. Ten adults from Port Allegany, Pennsylvania, are somewhat smaller, averaging as follows: total length, 167 mm; tail, 84.1 mm; hind foot, 20.7 mm. In Maine, this mouse is much grayer, appearing in the adult pelage similar to an immature individual of a New York mouse. There is seldom any indication of a dark dorsal stripe present, or if so, it is very faintly marked. Rarely a fawn-colored specimen is taken which resembles *P. leucopus*. Ten adult specimens from the Mount Katahdin region of Maine average as follows: total length, 184.4 (171–195) mm; tail, 91.2 (81–100) mm; hind foot, 20.8 (20–23) mm; weight, 16–24 grams. This mouse often occupies the same woods as *Peromyscus leucopus*. It may be distinguished by the longer, pencilled tail, the larger size, softer and grayer pelage, and larger ears. Fresh specimens show decided difference in the physiognomy of the facial region. The skull of *maniculatus* differs from *leucopus* in having longer nasals and less bulge of the maxillaries in front of the infraorbital

Figure 6.30.   Deer mice, *Peromyscus maniculatus gracilis* (above) and *P. m. bairdii.*

foramen, and in the parallel-sided palatine slits, which in *P. leucopus* tend to bow out in the middle.

DISTRIBUTION.   The larger, woodland long-tailed *Peromyscus maniculatus* occurs south into northern Wisconsin and Michigan, into New England and New York, and southward throughout the Allegheny Mountains (Figure 6.31).

HABITS.   The *maniculatus* group is one of the most widespread of all North American mammal species, with greatly variable populations occurring from Labrador to Tennessee and westward to Alaska and Mexico. These mice occupy diverse habitats: even in the eastern United States, the species resorts to quite distinctive quarters. In northern Maine *P. maniculatus* is at home in the dark spruce forests, whereas in New York, the species frequently chooses the hardwoods which are the resort of *Peromyscus leucopus,* while in the

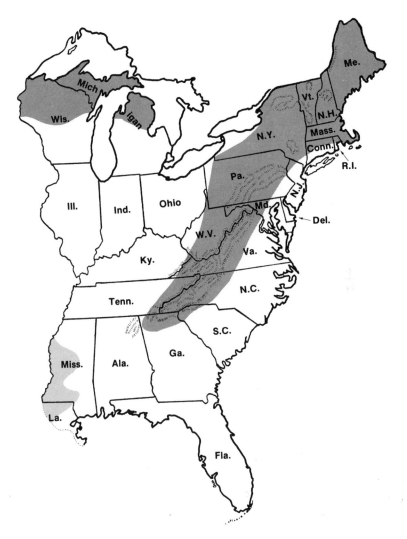

Figure 6.31. Distribution of the deer mouse, *Peromyscus maniculatus* (woodland forms), and the fulvous harvest mouse, *Reithrodontomys fulvescens* (lighter tint).

cloud-blanketed slopes of the Smoky Mountains and the higher Alleghenies we may look for it in mixed woods or conifers.

The woodland deer mice feed largely upon the seeds and mast of the forest or garner the ripe heads of grasses which abound in the clearings. They delight in the berries which serve their needs for a long period during the summer.

Numerous insects are eaten, and we have found in their stomachs the remains of little centipedes, green caterpillars, and even the remains of small birds and mice. As there is no thought of hibernation, they must cache a sizable store of nuts, seeds, and other delicacies for the long winter months. In summer their small internal cheek pouches may be found crammed with the seeds of blueberries, raspberries, and other winter staples.

The prairie deer mouse, *P. m. bairdii,* is superficially similar to *P. leucopus,* but its *open field habitat* and *short tail and ears* usually serve to distinguish it from other deer mice. The upperparts are brownish gray, mixed with darker hairs, brighter in winter, although considerable variation occurs within the same locality. Underparts are white, the basal gray of the underfur often conspicuous; ear dark brown, the margin edged with white; tail sharply bicolor, slightly pencilled. The short tail serves best to identify it. Average measurements of ten adults of *bairdii* from Michigan, Indiana, Illinois, and Kentucky are: total length, 152 (140-163) mm; tail, 63 (54-70) mm; hind foot, 19 (15.5-22) mm; weight, 16-26 grams.

The prairie deer mouse shuns the woodland. It inhabits the open fields and sandy areas, but reaches its greatest abundance in the corn, wheat, and soybean fields of the midwest, where it makes its home in the soil, not forsaking these fields even when the ground has been plowed. At that time, prairie deer mice feed on corn, soybeans, and grass seeds in the soil, while living in short burrows. In winter their tiny footprints can be found in these fields or in winter wheat fields, meandering from one burrow opening to another. We may also see the delicate tracery left by its footprints on the burning sand dunes which border the Great Lakes.

The prairie deer mouse occurs as far east as central New York (Tompkins County), the panhandle of West Virginia (Avalon, Ohio Counties), through most of Ohio, Indiana, and Illinois, southern Michigan and Wisconsin and south through most of Kentucky and western Tennessee. In recent years, with the cutting of the forest, this species has spread into northern Michigan.

Like other members of the genus, these mice are strictly nocturnal, seldom being taken during the daylight hours. Their lively nightly scampering and petty pilfering may drive the vacationer to distraction.

Breeding commences in the early spring and lasts well into the fall, although there is a slight let-up during the middle of the summer. From two to seven young (usually two to four in *bairdii*) are born in a warm nest situated in a stump, beneath a log, or in a hollow stub, although *bairdii* may seek no shelter other than a partially buried plank or log on the sandy beach, where shelter is at a preium. The young are blind, naked, and helpless at birth but grow rapidly, being weaned before they are three weeks old and ready to commence family duties of their own in another three weeks.

These mice climb with great agility and the woodland forms may take refuge in a squirrel nest or cavity many feet from the ground. Traps set well above the ground will take them. The prairie deer mouse will often occupy the corn shocks during autumn, constructing a cozy nest of husks and feasting on the kernels. It is a greater pest in this respect than the field mouse.

These little mice have well-marked homing behavior, for individuals removed two miles have returned to the point of capture within a few days.

Deer mice progress by a series of leaps, both the fore and hind feet leaving two well grouped pairs of tracks, quite unlike the trotting gait of the shrews and wood voles, but similar to tiny rabbit tracks. The tail leaves a conspicuous mark in the snow.

This species, like other small forest mammals, has periods of great abundance and relative scarcity. It is difficult to determine what occasions these more or less periodic fluctuations, but they must influence the habits, and possibly the numbers, of the great army of predatory mammals, hawks, and owls which seek to capture them at every opportunity.

Oldfield Mouse. *Peromyscus polionotus* Wagner

DESCRIPTION. This small fawn-colored *Peromyscus* occupies the open fields of cotton and corn and waste areas in the deep south and the beaches of eastern and western Florida. Individuals from inland populations of this mouse can be distinguished from other deer mice occurring in the southern states by their smaller size, soft brownish fawn color, and pale underparts, the hairs of which are slate gray at the base except on the chin and throat, where they are white to the base. Upperparts are uniform brownish fawn, or brownish gray, slightly darker on the middorsal line; sides of face somewhat brighter; ears dusky, pale-edged; tail bicolor, dusky above, white below; hind foot small, usually less than 19 mm; skull broad and short. Measurements of twenty adults from Georgia, Florida, and Alabama average: total length, 127.4 (122–138) mm; tail, 46.6 (40–51) mm; hind foot, 16.5 (15–18) mm. The darkest colored individuals of the species occur in northern Alabama, Georgia, and South Carolina.

The species varies greatly throughout its range, being darker in inland populations and lightest or white in populations occurring along shore on the white sandy beaches. The darkest individuals of all occur above the fall line in northern Georgia and Alabama. Several subspecies have been described in this species. South Carolina populations were studied by Albert Schwartz (1954) and gulf coastal populations by W. W. Bowen (1968). Bowen found reduced fertility between some series of adjacent populations, and also a possible case of circular overlap. Much more work is needed on this variable and highly interesting group.

Individuals from coastal Alabama and the western panhandle of Florida are grayish fawn color; white of underparts extensive; end of nose and narrow stripe extending nearly to interorbital region whitish; thighs whitish; tail all white except basal third, the upper part of which is dusky to pale grayish brown. Ten adults from St. Andrew's Point Peninsula and Cape San Blas, Bay, and Gulf Counties, Florida, are very pale and average total length, 126.4 mm; tail, 49.1 mm; hind foot, 18.2 mm. Average measurements of eight adults from the southeast coast of Florida are total length, 135.5 mm; tail, 52.4 mm; hind foot, 18.4 mm. Ten adults from Anastasia Island average: total length, 138.5 mm; tail, 53.5 mm; hind foot, 18.7 mm.

DISTRIBUTION.   *Peromyscus polionotus* ranges from eastern Alabama, Georgia, and western South Carolina south through much of Florida (Figure 6.32).

HABITS.   In the dry, neglected sandy fields and beaches of Georgia, Florida, and Alabama, a pale species of mouse has developed. These oldfield mice, or beach mice, as they are called, depending on the habitats which they occupy, are proficient burrowers, their tunnels marked by well defined mounds of earth at the entrance. At times these mounds are surprisingly large and suggest the work of a pocket gopher.

The burrow of this species normally slants from the entrance to varying depths, then levels off; at the end of the burrow a nest is constructed. A branch of the burrow normally extends directly above the nest to within an inch or two of the surface, and may serve as an escape exit if a marauding snake should enter the burrow. Digging into the burrow entrance with a shovel will often cause the mouse to "explode" from the sand and go dashing off. The burrows are sometimes closed, sometimes open; but all occupied burrows appear to be quickly closed by the mice if heavy rains threaten to flood out the inhabitants.

A. H. Howell (1921) has found *Peromyscus polionotus* abundant in eastern Alabama, where it appears to favor sandy fields with abundant cactus but sparse growth of grasses. These mice also venture into cotton and corn fields and occasionally occupy hedgerows or venture into open timber tracts.

The breeding habits are little known, although large numbers have been raised in the laboratory of the Scripps Institution of Oceanography for genetical studies by F. B. Sumner (1926), and W. W. Bowen has likewise raised hundreds of individuals in captivity.

This species and its related forms probably feed on the seeds of grasses, weeds, and grain; Howell has recorded the remains of blackberries in the stomachs of *polionotus*. *Leucocephalus* feeds on a wild pea (*Galactia*).

Carlyle Carr informed Hamilton that *polionotus,* at least in Florida, is at times a local nuisance to the planter, digging up melon seeds as soon as they are planted, and that its repeated depredations have necessitated three plantings of cantaloupes before a partial setting could be obtained.

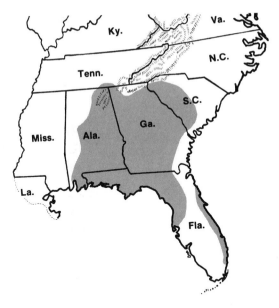

Figure 6.32.   Distribution of the oldfield mouse, *Peromyscus polionotus.*

White-Footed Mouse. *Peromyscus leucopus* (Rafinesque)

DESCRIPTION.   The tail of this medium-sized deer mouse is usually appreciably shorter than half the total length, and less hairy than the *maniculatus* group; the pencilled tuft is not prominent; the color is variable. In summer pelage the upperparts are grayish brown to dull orange brown, the middorsum darker, interpsersed with black-tipped hairs; the underparts are white, the bases of the belly hairs bluish gray but usually effectively concealed by the white tips; ears are grayish brown, the extreme margin pale white; preauricular tufts colored same as face (in *maniculatus* these are often whitish); tail dark brown above, white below, the line of demarcation usually not as sharply marked as in *maniculatus*. Winter pelage is grayer and the tail is more sharply bicolor, nose tip whitish. Immature mice are plumbeous gray above, belly and feet white. In the northern and Appalachian forests this species can easily be confused with the larger, long-tailed forest form of *P. maniculatus*. The best characters for distinguishing it are the tail length and skull characters. The tail is slightly *less* than half the total length in *leucopus*, whereas in *maniculatus* it is slightly *more* than half the total length of the animal. The skull differences are described under woodland *Peromyscus maniculatus*. *Peromyscus maniculatus bairdii* essentially never enters the woods, but in open areas it can be confused with *P. leucopus*. *Peromyscus maniculatus bairdii* is a smaller, shorter-tailed, grayer form, with a hind foot *usually measuring 18 mm or less*. The hind foot of *P. leucopus* is *usually 19 mm or more*. The tail in *P. leucopus* is usually just less than half the total length of the animal, whereas in *bairdii* it is usually considerably less. Average measurements of five specimens of *leucopus* from Easley, Pickens County, South Carolina, taken at an elevation of 1000 feet, are: total length, 155 mm; tail, 69 mm; hind foot, 20 mm. Nineteen adults from Tennessee average: total length, 165.4 (152–181) mm; tail, 71.8 (59–83) mm; hind foot, 20.1 (19.5–22) mm; weight, 15–25 grams.

In the northern half of the eastern United States, individuals are paler and somewhat larger; the pelage softer, particularly in winter specimens; tail more thickly haired. Individuals often have a pronounced orange cinnamon shade to the back, giving an appearance of "red mice." Long-tailed individuals in winter pelage are easily confused with woodland *P. maniculatus*; both forms often occur side by side in the same forest. The only certain measure of identity is skull comparisons by which the two can usually be separated with assurance. Measurements of thirty adults from western New York are: total length, 170.5 (157–189) mm; tail, 75.8 (60–92) mm; hind foot, 20.6 (18–23.3) mm; weight, 16–28 grams (Figure 6.33).

DISTRIBUTION.   The species occurs throughout most of the eastern United States, except for northern Maine and the southeast below the fall line (Figure 6.34).

Figure 6.33. Female white-footed mouse, *Peromyscus leucopus,* with three young.

Figure 6.34. Distribution of the white-footed mouse, *Peromyscus leucopus*.

HABITS.   This species is primarily a forest dweller, frequenting the dense woods or their borders, but also occurring abundantly in hedgerows and brushy areas. It less often ventures into open grassland and cultivated areas, except when hedgerows or woodlands are close by. It occasionally takes up a residence in the village; the first evidence of its presence in the house may be a boot half filled with cherry pits or hickory nuts in the cellar. It seldom ventures far from the woods, where its dainty tracks may be seen on the snow-covered forest floor on the coldest winter days.

The great black eyes, long sensitive vibrissae, and big ears are all stamps of a nocturnal existence, and it is rare to find this little mouse abroad other than at night. As dusk descends, it leaves the half-rotted stump or descends from some venerable beech whose cavity shelters it during the bright hours of daylight. In the winter the abandoned nest of some bird may be deftly capped over with leaves or thistledown, and in this warm retreat the deer mouse sleeps away the day. Or, for that matter, it may use a bird's nest to produce its young. Again, the home may be the deserted nest of a squirrel, high in the oak boughs, or a ball of dead leaves beneath the forest floor, for this little mouse is adaptable and makes the best of its environment.

In the most favorable habitat, it is one of our most abundant small mammals, every stump or hollow log seeming to harbor a mouse. In some years deer mice are incredibly abundant, and cause much mischief by entering the cabin or farmhouse.

In early spring the mice seek a mate, and three weeks after mating the tiny naked young are born in a warm nest of leaves and shredded bark. The gestation may be prolonged later in the summer if the female is nursing a previous litter. The young grow rapidly and when three weeks old they are clothed in a handsome gray coat which contrasts sharply with the cottony white underparts. If the parent be disturbed when nursing her young, she flees the nest with the entire litter clinging tenaciously to her teats. She need not run far, for every hole and cranny in her small range has been explored previously, and down one of these the parent and her burden disappear. Several litters are produced during the summer, and the young are able to breed when they are but two months old. Small wonder these mice sometimes attain such high populations. There is some indication that the females hold a territory during the season of reproduction, defending it against other mice.

During the winter deer mice are active, even in the severest weather, although several may huddle together in a warm nest in a bird house or woodpecker hole, remaining inactive for several days if stormy weather prevails. A few may actually hibernate. Brina Kessel and Hamilton, cleaning out starling nest boxes at Ithaca, New York, in the winter, recovered several deer mice that were torpid, exhibiting all the characteristics of a hibernating jumping mouse.

Homing behavior is well developed in this species. Hamilton has marked individuals and released them fully a mile from their home territory and taken them a few days later at the point of their first capture.

Like many other rodents, deer mice are omnivorous, although their chief food appears to be the little nutlets, berries, and seeds which abound in the forest. Black cherry seeds are a favorite food of this species, although they have to be removed from the pits. The seeds of jewelweed (*Impatiens*), known to many as Touch-me-not, taste like walnuts but have beautiful turquoise blue endosperm. That the seeds are heavily feasted upon can easily be observed by the many individuals of this species in late summer having turquoise-colored stomach contents which often are obvious even before stomachs are opened. The small internal cheek pouches often are crammed with such tiny seeds as blueberries, and they store the seeds of raspberries, jewelweed, and shadberries, wild cherry pits, and various species of viburnum. Several quarts of clover seed have been found in the galleries of a single one of these mice, and Hamilton once found nearly a peck of beechnuts which had been stored in the cavity of a large beech almost thirty feet above the ground. As fall approaches, the deer mice harvest all sorts of edibles, including quantities of basswood seeds, acorns, chestnuts, the seeds of conifers, and other items. These mice consume great quantities of insects during the summer months. Chief among these are groundbeetles and caterpillars. They eat centipedes, snails, an occasional small bird, and even other small mammals, including their own kind.

Enemies of the deer mice are legion; they number every owl, many predatory mammals, chief among which are the weasel and red fox, and quite probably snakes. The ubiquitous short-tailed shrew possibly overcomes these mice in their tunnels upon occasion, but it is hardly a match for *Peromyscus* elsewhere.

Cotton Mouse. *Peromyscus gossypinus* (LeConte)

DESCRIPTION.   This medium-to-large *Peromyscus* has a big hind foot; the tail is less than half the total length, not sharply bicolor, sparsely covered with short hairs; color is dark, the dorsal area broadly darkened; it is very similar to *Peromyscus leucopus* but heavier and considerably darker. Although occurring in the same localities as *P. leucopus* in the northern part of its range, and interbreeding in captivity, *gossypinus* in the field mates with its own kind, thus maintaining itself as a distinct species. In summer pelage, the upper parts are rufescent cinnamon, thickly interspersed with black hairs, lending a dark tone to the back, dark dorsal stripe broad but not well defined; nose, top of face, and head distinctly grayish, usually well marked off from brown cheeks; ears grayish brown, with no pale border; underparts dirty white, usually much duller than in *leucopus,* the region in front of shoulders and lower throat creamy; feet white; tail blackish brown above, dull white below but not sharply bicolor. Average measurements of eight adults from southern Georgia are: total length, 176 (164–192) mm; tail, 75.5 (71–87) mm; hind foot, 22 (21–22.5) mm.

Individuals from the general region of northern Georgia, Alabama, and

Tennessee are larger and paler than typical *gossypinus*. They are less dusky on the sides and the dorsal streak is less pronounced than in the coastal form. The underparts are likewise brighter, being creamy white. Average measurements of thirty adults from the mountains to the western lowlands of Tennessee are: total length, 185 (160–205) mm; tail, 80.1 (63–97) mm; hind foot, 23.3 (20–26) mm; weight, 25–39 grams. Cotton mice from peninsular Florida are not sharply marked off from typical *gossypinus*, but study of a sufficient series will show them consistently smaller and paler. The underparts are often creamy or yellowish white. Average measurements of twenty adults from Oak Lodge, east peninsula, Brevard County, Florida, are: total length, 181 mm; tail, 71.8 mm; hind foot, 21.5 mm. Those from the southwest portion of the peninsula are small and brownish, dorsum darker. External measurements of fifteen males and females from Collier County are 166 (152–189) mm, 71 (63–80) mm, and 22 (21–23) mm (Figure 6.35).

DISTRIBUTION.   The cotton mouse occurs in the southeast, north to southern Illinois, western Kentucky, and along the eastern half of South and North Carolina (Figure 6.36).

HABITS.   The cotton mouse occupies a variety of habitats, from the low swampland of the Atlantic coastal plain to the open woodlands of the Smoky Mountain foothills at an elevation of 2,000 feet. It occurs in the dense underbrush in the lowest and wettest parts of overflowed lands. Remington Kellogg (1939) mentions its preference for cliffs and rocky bluffs, especially caves and crevices. Cotton mice burrow in the dry ridges bordering the Louisiana bayous, making their nests wherever high water will not invade their tunnels. While its name suggests a habitat in cotton fields, Howell remarks that the

Figure 6.35.   Cotton mouse, *Peromyscus gossypinus.*

Figure 6.36. Distribution of the cotton mouse, *Peromyscus gossypinus.*

species is scarcely ever found in such places, unless the fields are on the borders of a timbered swamp. He states that *gossypinus* is a typical timber mouse and most abundant in the heavy swamps of the river bottoms. Major John LeConte, who described the species, mentioned its fondness for nesting under logs and the bark of decaying trees, making its home of cotton, frequently using more than a pound of this material for its purpose. In Florida these mice occur in the inland hammocks, in the salt savannahs, in piles of brush and rubbish about the cleared fields, and in the shaded retreats provided by the saw palmetto thickets.

The characters of this mouse are not too well defined in the conventional museum skins, and the skull, except for its slightly larger size, differs only in minor degree from that of *Peromyscus leucopus.* In the flesh, however, these related species look very different and are easily distinguished.

Like its white-footed relative of the north, the cotton mouse inhabits the haunts of man, invading the house or camp for shelter and partaking of the crumbs and stores of human occupants. Francis Harper reports that a trapper of the Okefinokee Swamp whose provisions were being raided set a single trap and therewith captured twenty-two cotton mice during a winter evening.

Harper (1927) himself had a somewhat similar experience when he set a single trap and took five mice during the course of an evening.

Little is known of the feeding habits of the cotton mouse, but there is no reason to suppose that their food differs from that of other deer mice. Outram Bangs records their feeding largely on the seeds of sea oats along the Florida beaches.

The breeding season occupies a long period, and, as with some other small mammals of the southern states, may persist through the winter months. Females collected in early March in eastern Tennessee have contained embryos, and a Georgia specimen captured in late December contained one large embryo. The litter is apparently smaller than with the northern forms, but the longer breeding season, which probably results in a greater number of litters, enables these mice to maintain themselves in spite of numerous predators. The enemies of the cotton mouse must include the ubiquitous barred owls whose weird calls echo from every southern swamp, certain hawks, the abundant blacksnakes and chicken snakes (A. H. Howell records finding a cotton mouse in the stomach of an Alabama rattlesnake), and of course skunks, weasels, and foxes. These all serve to reduce the cotton mouse population.

Peromyscus leucopus and *P. gossypinus* are so similar in structure and seemingly in habits that it would seem that they would be in rather direct competition when they occur together. A careful comparative life history study of the two where they occur together would be of great interest and would perhaps indicate some differential behavior patterns allowing them to live together in apparent harmony.

Florida Mouse. *Peromyscus floridanus* (Chapman)

DESCRIPTION.   In this large *Peromyscus,* which has big, nearly naked ears, a relatively short tail, and a very large hind foot, the *plantar tubercles number but five instead of six* as in other eastern species. The very soft and silky fur is bright ochraceous buff on the sides and white below. Fresh specimens can be recognized at once by the number of the plantar tubercles and the very large hind foot. Average measurements of ten adults are: total length, 196 (186–207) mm; tail, 88 (85–94) mm; hind foot, 26.4 (24–29) mm. See Figure 6.37.

DISTRIBUTION.   The Florida mouse is found in the central part of peninsular Florida, from coast to coast (Figure 6.38).

HABITS.   Outram Bangs (1899), who contributed so much to North American mammalogy, wrote of this species: "The big-eared Florida deermouse is common in all suitable places throughout peninsular Florida—It

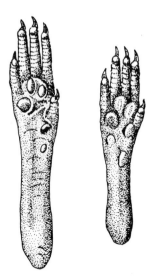

Figure 6.37. Sole of hindfoot of *Peromyscus floridanus* compared with hindfoot of *Peromyscus leucopus.*

lives only in the higher sandy ridges, where there is plenty of black-jack oak and turkey oak, and where the bare white sand is in places covered by scattered patches of scrub palmetto. It is the characteristic small animal of such places, commonly known as 'black-jack ridges,' and I have never found it elsewhere.''

The name ''gopher mouse'' has been applied to this species because it frequents the burrows of the gopher turtle, *Gopherus polyphemus.* The mice share the burrows with the turtles, and possibly make small holes into the side of the burrow to serve as nests.

One of the greatest difficulties which beset the southern collector is the countless millions of ants occupying all the dry ridges. These insects riddle the stumps, tunnel the soil, and invade the very burrows of small mammals. They eat the bait from the pan as soon as the trap is set on the ground, or, if

Figure 6.38. Distribution of the Florida mouse, *Peromyscus floridanus.*

the collector is fortunate enough to take a specimen, it is often hopelessly mutilated before it can be removed. The repeated predations of these innumerable hordes serve to discourage the small-mammal collector, and even if he persists in his efforts, the results are seldom comparable to the effort involved.

Golden Mouse. *Ochrotomys nuttalli* (Harlan)

DESCRIPTION.   The handsome little golden mouse is characterized thus: soft, thick pelage and heavily furred underparts; rich, tawny ochraceous uniform upperparts, slightly paler on sides, head, and ears; no eye ring; feet and underparts creamy white, the latter often strongly suffused with ochraceous, particularly on the abdomen.

Average measurements of ten adults of both sexes from the Smoky Mountains are: total length, 176 mm; tail, 85 mm; hind foot, 19.7 mm; weight of adult: 20–26 grams (Figure 6.39).

In the southern parts of its range the species averages smaller. Measurements of ten adult Florida mice average: total length, 158 mm; tail, 71; hind foot, 18.4 mm. Individuals from Florida are yellowish brown dorsally. There is a lack of black guard hairs on the posterior dorsum, giving an orange or golden-red pelage. The belly is cream buff with a yellowish overtone

Figure 6.39.   Golden mouse, *Ochrotomys nuttalli.* Photo by Roger Barbour.

extending onto the region of the jaws. East central Virginia individuals have darker underparts, with more black; ears dusky instead of ochraceous-tawny; tail fuscous above, white below. Young are without the tawny coloration of typical *nuttalli*. In appearance, adults are quite similar to the reddish phase of *Peromyscus leucopus*. Individuals from the western part of our area are rather small and pale.

DISTRIBUTION.   The golden mouse occurs from western Virginia and eastern Kentucky south through North Carolina to the middle of the Florida Peninsula, westward to Louisiana and southern Illinois (Figure 6.40).

HABITS.   The dainty little golden mice are widely distributed, and, as one might expect, are found in a variety of habitats over their extensive range. In the mountains of Virginia, Kentucky, and Tennessee, they occupy the pine and greenbrier thickets, the boulder-strewn slopes of dense hemlock forests, and the borders of broom sedge fields. In the southern swamps this beautiful little mouse is not uncommon; here it makes its little nest in the strands of Spanish moss which drape the live oaks and bushes of the low ground. In Alabama it occupies the canebrakes and swampy woodland, or, more rarely, it occurs in dry, thickety flatwoods or among the hills.

The golden mouse is largely arboreal, and often climbs to a height of thirty or more feet. It constructs its cleverly woven nest at varying heights in a bush or crotch of a tree. The mouse runs among the limbs with amazing celerity, the tail acting as a balance and even as a prehensile organ for clinging to branches when the footing becomes hazardous. The feet are smaller than those of other members of comparable size in the genus, and this may be a scansorial adaptation.

Roger Barbour informed Hamilton that at Morehead, in eastern Kentucky, where these mice are common, they build bulky nests of dead leaves and pine needles lined with finely shredded bark. The nests vary considerably in size, from that of a football to some scarcely larger than a baseball. The larger nests are occupied by several mice; occasionally one finds as many as eight in a nest. These nests are placed at various heights from a few inches to ten or more feet from the ground. The larger nests provide quarters during the winter and are placed in bushes, the branches of fallen trees, or the notch of a standing oak or hickory. Some of the nests, usually smaller ones, are used mainly as feeding shelters. The females often construct their breeding nests in impenetrable thickets of greenbrier or honeysuckle. Occasionally nests are placed on the ground, under the mantle of a protecting log or stump.

Little is known of the home life or reproductive habits of the species. Young mice and pregnant females have been found from early April throughout the summer months in the northern parts of the range. In Louisiana the species may breed at any time during the year, although more young are

Figure 6.40. Distribution of the golden mouse, *Ochrotomys nuttalli*.

produced in fall and early winter than at other times. The young are blind and naked at birth, have dark brown hair over the back and hips by the fifth or sixth day; the eyes open between the eleventh and fifteenth days.

Probably the feeding habits are similar to those of *Peromyscus*, but no studies have been made on this subject.

The golden mouse is relatively rare in collections, but it is by no means a scarce creature. Those who are acquainted with its habitat have succeeded in taking large numbers in a short period.

Hispid Cotton Rat. *Sigmodon hispidus* Say and Ord

DESCRIPTION.   A small robust rat, the cotton rat has a medium long tail and grizzled long coarse pelage. Pelage above is grizzled buff and black on hair tips, bases plumbeous; sides paler, tawny or yellowish brown; ears black, feet varying from gray to dark brown; undersides pale gray or buffy; tail scaly, very scantily haired, dusky above, paler below but not bicolor. Measurements of eight adults from North Carolina and Georgia average: total length, 268 (215–285) mm; tail, 108.5 (75–111) mm; hind foot, 30.6 (26–33) mm; weight, 80–120 grams (Figure 6.41).

Individuals from peninsular Florida are smaller, the brown of the upperparts much reduced and replaced by gray. The gray or yellowish gray and black tipped hairs give a characteristic pepper and salt appearance; underparts grayish white. Measurements of four adults from Fort Myers, Florida, average: total length, 239 mm; tail, 95 mm; hind foot, 29.5 mm. Extreme southern Florida individuals have a browner, less gray dorsal pelage; the rump is cinnamon rufous; tail almost uniform dull black, but slightly paler below. Five adults in the United States National Museum, measured by A. H. Howell, average as follows: total length, 278 mm; tail, 197 mm; hind foot, 32 mm.

DISTRIBUTION.   The hispid cotton rat occurs in southern Virginia, southern Florida, west to Tennessee and Louisiana (Figure 6.42).

HABITS.   Every geographic region has certain characteristic species, some more abundant than others, and often one species dominates over all the rest. In the waste fields and lush meadows of the north the field mouse reigns, but in the south the cotton rat is king. Every field of broom sedge, every roadside ditch, even the open glades of the forest to elevations of 1,700 feet

Figure 6.41. Hispid cotton rat, *Sigmodon hispidus.*

Figure 6.42. Distribution of the Hispid cotton rat, *Sigmodon hispidus.*

have their quota. The canopy of grasses and weeds covers a multiple of well-defined trails, and where the cover is thin these highways can be seen for some distance.

While it is evident that the cotton rat population fluctuates from year to year, there is hardly a season in which they do not outnumber all other small southern mammals. It is usually the most abundant species on the farmlands of the south.

Long shallow runways are constructed, in which are chambers for the reception of the small nest built of grass and available trash. The cotton rat is active both day and night. Occasionally a tract which harbors a number of these rats is burned over, and as the fire reaches their retreats the panicky rodents boil out in endless numbers. Many other species of small mammals also use cotton rat burrows.

Few mammals are more prolific. Breeding commences in the late winter and continues through October or November—perhaps at times throughout the entire year. Gestation is twenty-two days. The young may number one to twelve, but there are usually four to eight in a litter. The young weigh 6.5 to 8 grams at birth. The eyes open during the first day, and at the age of five or six days the young begin to leave the nest and fend for themselves. At that time they weigh only 10 to 20 grams. This makes the species one of the most precocious of all rodents. The parents then immediately produce another litter, thus adding to the great number of individuals often found of this species. Of six pregnant cotton rats collected in southern Arizona, the embryo number averaged twelve. Those which we examined from southern Florida in late March were not breeding.

Flesh and grass share in the dietary of this rat. The burrows are kept shorn of new growth, and little piles of cut sedges and various grasses appear at irregular intervals in the surface runways. Bachman wrote many years ago that this species was very destructive to quail, and more recently Herbert Stoddard (1932) has confirmed this, finding that the cotton rat, at least in Georgia, is a very destructive enemy of the bobwhite, destroying considerable numbers of eggs and chicks. The cotton rat also travels the ditches, feeding on crayfish and the little fiddler crabs, and we suspect its diet is occasionally varied with a juicy insect.

The cotton rat is the mainstay of the predatory birds, mammals, and snakes of the Southland. Before it is fully dark, the barred owl is abroad, in the swamps and cypress bogs; the gray fox and raccoon seek it; among the canebrakes and the cabbage palms, rattlers, copperheads, and many other snakes strike it down.

Bachman wrote of this species: "It is a resident rather of hedges, ditches, and deserted old fields, than of gardens or cultivated grounds; it occasions very little injury to the planter."

Perhaps this was because there was enough for all in the early days, or perhaps because the planter did not cultivate extensive truck crops for the northern markets in 1840. Today it is a pest of major importance to agriculture in many parts of the south. It may cause 50 percent loss to sugarcane by cutting the stalks off close to the ground. In the important truck-farming sections south of Miami, Florida, during 1931, the loss to truck crops through these rats was placed at $150,000 in spite of systematic control measures. Three or four plantings of squash seeds are often necessary before a partial stand of vines can be produced. The rats are particularly fond of sweet potatoes and are attracted from long distances to a field of these. The systematic poisoning of a one-acre patch resulted in the destruction of 513 cotton rats.

Eastern Wood Rat. *Neotoma floridana* (Ord)
Synonym: *Neotoma magister*

DESCRIPTION. This large ratlike rodent, which superficially resembles the Norway rat, may be distinguished from it by the large naked ears, very long black or white vibrissae, and the hairy tail. Northern wood rats have a distinctly bicolor tail. Pelage is buffy gray above, the hairs darker in the midline, sides buffy, head gray; tail dark gray above, white below; underparts white, with occasional buffy patches on breast; feet white. The winter pelage is slightly darker and longer. The crowns of the molars are flat with the enamel thrown into prismatic folds, thus differing from the tuberculate molars of *Rattus*. Measurements of ten adults from New York, Pennsylvania, and West Virginia average: total length, 423 (405–441) mm; tail, 186 (170–200) mm; hind foot, 43.5 (40–46) mm; weight, 370–455 grams. In the southern parts of its range this species has shorter fur, and the tail is scant-haired and not bicolor.

Wood rats from Louisiana and southwestern Mississippi are slightly larger than typical *floridana* (basilar length of skull averaging 42 mm), color more reddish, the upperparts cinnamon to dark ochraceous buff, brownish drab of middle of face darker than cheeks. Dusky hairs on top of head and back give this race a more intense and richer coloration. The underparts and feet are whitish; tail blackish above, only slightly paler below. Individuals from Illinois are similar to the preceding, but lack the reddish tone; color grayer and tail bicolor, blackish above, dull white below. Upperparts are dull buffy mixed with black; face grayish, outer sides of hind legs brownish; underparts white, the fur pure white to roots. Eight adults from Wolf Lake, Illinois, average: total length, 403 mm; tail, 195 mm; hind foot, 38 mm. Measurements of twenty-four adults from Georgia and Florida average: total length, 393 (362–409) mm; tail, 181 (166–189) mm; hind foot, 37.5 (36–40) mm; weight, 200–275 grams (Figure 6.43).

DISTRIBUTION.   Wood rats occur from extreme southern Illinois, Indiana, Ohio, and New York, south through the Alleghenies to Louisiana, Alabama, Georgia, and central Florida. They are absent along the coast from New

Figure 6.43.   Eastern wood rat, *Neotoma floridana*.

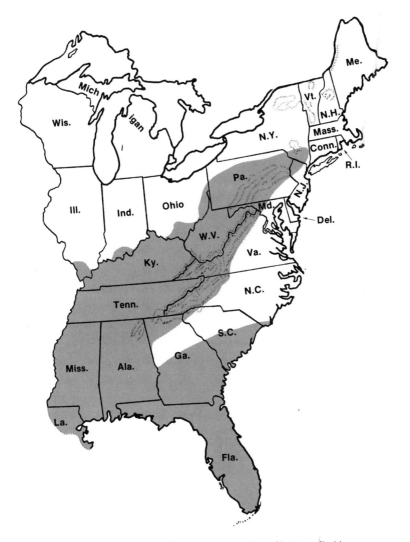

Figure 6.44. Distribution of the eastern wood rat, *Neotoma floridana.*

Jersey through central South Carolina and part of east central Georgia (Figure 6.44).

HABITS. The extensive range of this species accounts for its variable habits and mode of life. Rocky cliffs, caves, and fissures or tumbled boulders on the sides of mountains are the preferred habitat. In the rock slides and extensive bare patches of the Alleghenies, tumbled boulders form crevasses and extensive galleries, almost devoid of vegetation. Here the wood rats scurry by night, their presence made evident by the piles of droppings and trash which they lug to the dens and ledges. Such conditions are not found in southeastern United States, and here the rats live quite a different life. In much of Alabama, these rats appear to favor the osage orange hedges, where they build large nests in the branches of trees. In Georgia and Florida it is found in low wet ground in hammocks and swamps, where it lives in hollow trees or holes in the ground, or constructs large nests of sticks, leaves, and rubbish along the banks of streams in dense tangles of cabbage palmetto, or rarely in trees. These lodges are rather compactly constructed and sufficiently well knit to be waterproof. Within these retreats, nests of shredded bark are constructed, these opening to the outside by one or two entrances, although the exits may lead into the earth and extend several feet from the lodge before reappearing on the surface. In the north, the nest is a bulky structure built on a rock shelf or the level floor of a cave and well protected from the elements. It is open at the top, not unlike the nests of many birds. Many of those we have seen have been made of red cedar twigs and lined with shredded cedar bark, but any fine material available is undoubtedly used. Indeed, in Indiana we have often seen entire banks, above and below wood rat dens, strewn with cedar twigs. One nest was found in Indiana in an old building. It was entirely of trash and was on a seven-foot-high shelf. It covered the entire shelf, which was about four by five feet. There is some evidence that these rats colonize, for Francis Harper recounted the capture of three adult females at a group of three nests in the space of two days.

The wood rat's presence may go unsuspected for a long time in areas where they are not at all uncommon. Perhaps their seeming absence is due to the nature of the habitat which they select, for the inhospitable cliffs invite few hikers, and the nocturnal habits of these handsome rodents make them appear rare. They skip over the rocks with great celerity, visiting the brush and open thickets in search of food; the foreign material in their mountain nests suggests that they make extensive journeys into the bottomlands as well.

They are occasionally abroad on dull days, but the major share of their activity occurs in the hours of darkness, and usually the night is half gone before the peak of activity is reached. The wood rat does not stir much on stormy nights, and such movements as they make at these times are in the recesses of the caves or deep fissures which they occupy.

One of the characteristic traits of the wood rat is to lug all sorts of rubbish to the den site. This booty may include bone scraps, leaves, bits of wood, tobacco tins, shotgun shells, cast-off clothing, and the refuse from a camp site. Mixed with these may be a handful or two of the dried scats of the rat, but usually this creature repairs to some common site to deposit its oval pellets. These pellet accumulations may total several quarts and are infallible evidence that *Neotoma* is about.

When not breeding, these rats are aggressive creatures, fighting over their food or nest sites, and continually chasing one another. One seldom captures an adult which is unscarred; it may have torn ears or fresh skin wounds or may even lack a piece of the tail.

In the north, the breeding season continues from early spring until mid-fall, and two (perhaps three) broods are raised. The litter usually numbers one to three, most often two. The gestation period is thought to be between thirty and thirty-six days. As with other mice, the young are born in an immature and quite helpless condition. Their eyes open in the third week and another week has passed before they are fully weaned. In the south, breeding may occur irregularly through the year.

The diet of the wood rat is a varied one, and includes most of the plants in its domain. Fruits and berries, including dogwood, blackberries, mountain ash, wild cherries, and shadberries, the fruits and stalks of pokeweed and sassafras, fungi, ferns, rhododendron, and a host of other plants are all collected in the den. These are often left in a green condition on the rocks, which may indicate a hay-making practice similar to that of the pika. A. H. Howell (1921) believes them to feed in large measure on hickory nuts in Alabama, since he found a great accumulation of shells about the cliffs occupied by the rats. He likewise found pawpaw seeds about the dens. The few stomachs examined from Indiana wood rats all were entirely full of various kinds of green vegetation.

We can say from personal experience that the meat of this species is excellent.

The enemies of the cave rat are numerous; in spite of the protection of its rocky fortress it is frequently overcome by alert predators. Although it lives in company with the timber rattlesnake and the copperhead, Luther Hook, who has examined many of these snakes in the wood rat country of New York and New Jersey, has never found rat remains in their stomachs. Wildcats, foxes, weasels, and owls haunt the rocky cliffs and undoubtedly take a number. Probably the most feared enemy is the great horned owl, who can trespass on noiseless wings into the very stronghold of *Neotoma*, but it would seem likely that the large colubrine snakes take a toll.

The mountain people tell us that this big rat sometimes descends from the hills to capture young chickens. It can hardly cause much loss in this fashion.

Gapper's Red-Backed Mouse. *Clethrionomys gapperi*
(Vigors)

DESCRIPTION. This medium-sized vole has small eyes and relatively prominent ears which reach above the fur; a short tail; the long pelage is characterized by a broad rusty or reddish dorsal band, grading into buffy on the sides, and pale gray underparts. In winter pelage the dorsum is rufous chestnut, the hair tips sprinkled with black; facial portion of head and sides ochraceous buff; belly buffy white to pale gray, but with hairs dark plumbeous at the base; feet gray; tail slightly bicolor, yellowish brown to dark brown above, grading into buffy white below. The summer pelage is darker and more subdued. A pronounced color phase is frequently found in this species, the reddish dorsal stripe being replaced by a brown or grayish black pattern, so that the animal may be mistaken for a field mouse at first glance. Immature mice do not exhibit such a prominent reddish dorsal stripe. Average measurements of twenty adults from western New York and northern Pennsylvania (Allegheny County) are: total length, 138.5 (123–155) mm; tail, 38 (34–44) mm; hind foot, 19 (17–20.5) mm; weight, 20–28 grams (Figure 6.45). Northern New England individuals differ from typical *gapperi* in their strongly ochraceous tints and almost entire lack of red. They are slightly larger than the typical form, with longer, softer fur, and the black-tipped hairs on the back and sides are never conspicuous. Average measurements of sixteen adults from Mount Washington, New Hampshire, are: total length, 152 mm; tail, 40 mm; hind foot, 19.2 mm. Individuals from the southern Appalachians are decidedly larger than their northern relatives. The dull chestnut dorsal stripe is very broad, spreading out laterally and fading insensibly into the fulvous suffusion of the sides, which in turn encroaches on the white belly. The belly is usually strongly washed with ochraceous. Twenty adults

Figure 6.45. Gapper's red-backed mouse, *Clethrionomys gapperi.*

from the Smoky Mountains average: total length, 146 mm; tail, 44.7 mm; hind foot, 20 mm.

The red-backed mouse of mountainous southeastern Kentucky and southwestern Virginia is dark dorsally and duller buff on the sides. Average measurements of eight adults from Black Mountain, near Lynch, Harlan County, Kentucky, are as follows: total length, 153 mm; tail, 38 mm; hind foot, 19.5 mm. In the cedar swamps and sphagnum bogs of southern New Jersey, *C. gapperi* is pronouncedly darker than typical *gapperi,* with a darker tail pencil. Unlike the typical form, the upper body colors reach down on the sides and are abruptly separated from the whitish underparts. The hind feet are dusky gray. It is slightly smaller. Ten adults from Ocean, Atlantic, Cape May, and Gloucester Counties, New Jersey, average: total length, 136 (121–150) mm; tail, 36.8 (33–40) mm; hind foot, 18.4 (18–20) mm; weight, 118–30 grams.

DISTRIBUTION.   The red-backed mouse occurs in the forested regions of eastern United States, southward in the mountains to North Carolina and Tennessee (Figure 6.46). In the southern part of its range it is restricted to isolated bogs or mountain tops.

HABITS.   The cool shaded woods and moss-covered boulders of the eastern mountains are the haunts of these little red-backed wood mice. The treeless alpine summit of bleak Mount Washington, where they are one of the most abundant mammals, the cold damp sphagnum bogs of New Jersey, the spruce-covered cloud-hidden summits of the towering Smoky Mountains and the aspen meadows of northern Wisconsin all harbor these handsome little rodents. The red-backed vole seldom ventures from the forest, although it may be found in the grassy clearings of wooded regions, or on the treeless "balds" of the North Carolina mountains.

Little pretense is made of preparing elaborate tunnels similar to those of its cousin, the field mouse. In company with other woodland species, it occupies the burrows made by moles and shrews, or wanders about under the shelter of fallen leaves. Unlike the deer mouse with its hopping gait, this species progresses by trotting. It is an agile climber, scampering over the fallen logs or running up the windfalls to feed on the lichens and succulent edibles that carpet the rotting limbs. We have several times seen it about in broad daylight, scampering over the forest floor, its coat blending admirably with the forest carpet of dead beech leaves.

The red-back feeds on a variety of green buds, succulent roots, and fungi. Cut leaves of wild lily-of-the-valley (*Maianthemum*), bunchberry (*Cornus canadensis*), and other boreal plants here and there in the runs and among the rocks where this species abounds indicates its predilection for green vegetation. We have examined many stomachs but seldom found insect remains, although the blackberries and raspberries which grow in the open woodland

Figure 6.46. Distribution of Gapper's red-backed mouse, *Clethrionomys gapperi.*

glades and the fallen blueberries provide many meals. As this little wood mouse is active even during the most severe winter weather, it must lay aside large stores of food during times of plenty. During still-hunting in November, Hamilton saw one of these mice busy cutting the petioles from the leaves of wintergreen and carrying them to an underground cache. While hunting for deer, a friend once saw a dozen of these mice in an afternoon, all busily searching for beechnuts, which they industriously carried to some storage place.

Breeding occupies a long season, commencing in the late winter or very early spring and continuing late in the fall. Hamilton has found large embryos in New York specimens taken in early December. The young vary in number from two to eight. They are born after a gestation of seventeen to nineteen days. The young mice grow rapidly and are capable of breeding during the season in which they are born. One litter follows another through the warmer months. Following a good breeding season, when predators are few and conditions are generally favorable for an increase, the woods fairly swarm with them. In the northern part of their range, red-backed mice are the most numerous of all small mammals. Hamilton saw six running about near his temporary camp in a New Brunswick spruce forest, while in the White Mountains of New Hampshire and the Adirondacks of New York they are at times incredibly abundant. One trap will often serve to catch several in a single night.

Occasionally these mice prove a nuisance to plantings of ornamentals, and Merriam relates how they girdle trees a foot through to a height of three or four feet above the ground. The damage to deciduous trees may thus be great, but it is usually of a local nature. Merriam (1884) remarks on the tender and well-flavored flesh of this mouse. Hamilton has sampled the meat but found the picking too lean to pass judgment on its merit.

Key to Voles of the Genus Microtus

A. Tail about equal to hind foot; fur very fine and molelike (tending to lie either way); skull *very* similar to that of *M. ochrogaster* . . . .*Microtus pinetorum*
AA. Tail longer than hind foot. Fur coarser.
    B. Yellowish colored nose; third upper molar with six triangles . . . .
    . . . .*M. chrotorrhinus*
    BB. Nose not yellowish. Third upper molar with two or four triangles.
        C. Ventral fur silvery; tail usually much more than twice length of hind
           foot. Third upper molar with four triangles . . . . .*M. pennsylvanicus*
        CC. Ventral fur usually buff colored. Tail usually about twice length of
           hind foot. Third upper molar with two (three) triangles . . . .
        . . . .*M. ochrogaster*

Meadow Vole. *Microtus pennsylvanicus* (Ord)

DESCRIPTION. This relatively large, robust vole has a rather short tail, which is usually twice as long or longer than the hind foot. The fur is dense and soft, overlaid with a few coarser hairs. The short rounded ears are prominent, but may be well hidden in the winter pelage. The hind feet have six tubercles on the soles (in *M. ochrogaster* there are but five tubercles). The last upper molar has four intermediate triangles and a posterior loop; that of *M. ochrogaster* has two intermediate triangles and a posterior loop. The color of the upperparts in summer pelage is dull chestnut brown; a few individuals are bright chestnut; the dorsal pelage is interspersed with numerous black hairs; the belly is gray, occasionally tinged with buff as in *M. ochrogaster*. Feet and tail are dusky above, paler below, but the tail is not sharply bicolor. Winter pelage is darker with more gray. Immature specimens are much darker, with black feet and tail. Specimens from the southern part of the range are larger and much darker. Specimens in a large series from Chincoteague, Virginia, are notably darker than specimens from New York and Michigan, being almost black on the rump. Measurements of fifty adults from central New York average: total length, 167 (149–196) mm; tail, 42 (32–57) mm; hind foot, 21 (19–23) mm; weight, 25.60 grams. Twelve adults from Chincoteague, Virginia, average: total length, 171 mm; tail, 43.5 mm; hind foot, 23 mm (Figure 6.47).

On the dark tidal flats of extreme southeastern coastal Virginia and northeastern North Carolina, occurs a large dark race with large hind feet (originally described as *M. p. nigrans*). The color of the upperparts in summer pelage is dull brown, with many black hairs; rump very dark, almost black; underparts smoky gray, with buffy or cinnamon markings in some specimens. Winter pelage is much darker, almost black above; feet and tail blackish. Young are sooty black all over. Measurements of three adults from southeastern coastal Virginia average: total length, 165 mm; tail, 48 mm; hind foot, 23 mm.

Figure 6.47. Meadow vole, *Microtus pennsylvanicus*.

Figure 6.48.   Distribution of the meadow vole, *Microtus pennsylvanicus.*

DISTRIBUTION.   The meadow vole or field mouse occurs from Maine to
South Carolina, and westward through Wisconsin and central Kentucky and
northern Illinois. This species has not definitely been recorded from Tennessee
(Figure 6.48).

HABITS.   Low meadows and swampy pastures, fields with a protecting
cover of dead grass and herbs, and the salt meadows of the coast are all
favored resorts of this mouse. Less often it is found in open glades in the

woods, where the sun encourages the growth of rank vegetation and herbaceous annuals. Every conceivable niche supports a few; the adaptable creature is thus found over a very extensive range. Indeed, few mammals are so widespread or found in such a wide variety of habitats.

In the meadows it makes numerous trails about the width of a garden hose. These are kept bare and smooth by the sharp teeth of the mice, which soon level any green sprout which forces the soil of the runway. At irregular intervals great masses of the little brownish green pellets are found, communal toilets for these cleanly little beasts. The field mouse is a social creature, although each individual appears to have a well established range, seldom encompassing an area larger than a tennis court. Their ranges overlap and as a consequence many may inhabit a very small area. Meadow mice are cyclic rodents, their populations rising and falling rather regularly. Every three or four years the population becomes high, occasionally numbering 200 mice or more to an acre in favored localities, and then it is quickly levelled, apparently through psychological and physiological responses to overpopulation (Hamilton 1937).

Field mice are about by day and night, but they seem to favor the early morning hours and those of the late afternoon.

The nests of dried grasses are quickly but strongly made. These are built either directly on the surface of the ground or at the end of shallow burrows. The heat from the occupants melts away the snow during the winter months so that a chimney is formed, revealing the compact nest below. In swamps and other wet places the nest may be placed in the center of a grass tussock, well removed from the threat of high water, forming a wet but by no means impassable barrier to the feeding stations.

Winter holds no fear for these little rodents. Although they seldom store any quantity of food, the tender bark of trees, the blanched shoots of grasses, and various seeds and hardy rootstocks provide abundant food even during the most severe winter weather. As the spring thaws level the snow banks, abundant evidence of their subnivean activity is revealed.

Among the many prolific mammals, the field mouse is champion. One large litter follows another in rapid succession, until it seems that the countryside should boil over with these little mice. A captive female observed by Vernon Bailey (1924) produced seventeen litters in a single year, and one of her daughters from a first litter had produced thirteen families of her own before she had reached her first birthday! No other mammal can challenge such fecundity. The gestation period occupies three weeks, but the three to ten young, although born in an immature state, are weaned when twelve days old and the females are ready to breed when they have attained the ripe old age of three weeks. The breeding season is a long one, continuing from late March into November, and in some years they breed throughout the winter.

It would be easier to list the vegetative foods that are not eaten by *Mi-*

crotus than to list those that it does consume. Grasses of all kinds and their maturing seeds, fleshy rootstocks, and the tender bark of even sizable trees are all relished. In spring it will cut down dwarf willows and eat the fruiting bodies. In the fields of closely growing timothy and grains, the stalks are so closely crowded that the mice must cut them into match-sized sections before the prized heads can be reached. This accounts for the little criss-cross patches of cured grasses the puzzled farmer observes so frequently in his fields. In the salt meadows, various sedges and even the tiny littoral life is eagerly taken, but the field mouse, unlike many of its kin, appears seldom to eat insects. Its appetite is prodigious, a mouse often eating green food in excess of its weight in twenty-four hours.

Of all our small mammals, none has such a long list of enemies. In the air above, the woods and meadows, and even in the water lurk many foes, always ready to snuff its life in one savage rush of wings, feet, or fins. Were it not for the widespread distribution and large numbers of this rodent, many predators would be hard-pressed to maintain themselves.

Field mice are of great economic significance. By girdling fruit trees and nursery stock they cause considerable monetary loss to the horticulturist. Their constant pilfering of forage crops, while difficult to measure, is in the aggregate a very great loss to the agriculturist. They do serve a useful function in providing predatory birds and mammals with an abundant source of food.

Rock Vole. *Microtus chrotorrhinus* (Miller)

DESCRIPTION.   The general appearance of the rock vole is similar to that of *Microtus pennsylvanicus,* but this species can at once be distinguished in the field by the orange or saffron nose. It appears similar in body characters to the field mouse, but differs in the larger ear, smaller hind foot, and orange wash on the rump. In summer pelage the upperparts are grizzled brown or bister, mixed with black; facial region and to a lesser extent below ears, orange rufous or saffron; the same color faintly defined on the rump; belly silvery gray to plumbeous; feet silvery gray above; tail sepia above, pale below. Average measurements of five adults from Mount Washington, New Hampshire, the type locality, are: total length, 168.5 mm; tail, 48 mm; hind foot, 19.5 mm; weight, 30–40 grams.

Southern Appalachian individuals (originally described as subspecies *carolinensis*) differ in having a larger skull, longer upper tooth row, and heavier zygomata. Upperparts are dark, blackish bister; sides slightly blacker; nose to eyes deep orange rufous; small spot of rufous back of ear but this is not constant; underparts dark plumbeous; tail above colored like back, paler below. Measurements of the skull, with those of typical *chrotorrhinus* following in parenthesis, are: occipitonasal length, 26.5 (25.8) mm; greatest zygomatic breadth, 15.1 (14.3) mm; upper molar series, 7.1 (6.5) mm; weight, 26–47 grams.

DISTRIBUTION.   The rock vole occurs from Labrador to North Carolina, but its distribution is spotty and it occurs only in the most suitable localities (Figure 6.49).

HABITS.   Among the shaded retreats formed by moss-covered boulders of northern mountains or about the logs of the high humid southern mountains, this little saffron-snouted mouse passes its life, seldom seen even by the professional mammalogist. Collectors have taken specimens in the rock slides of Mt. Washington, well above the timber line. It is not uncommon in the Adirondack and Catskill Mountains of New York, nor in the cool forests of northeastern Pennsylvania, but other species usually greatly outnumber the rock vole and it is rare in collections. Whitaker and Robert Fisher found it to be abundant on Slide Mountain in the Catskills of New York, and took twenty-five individuals relatively easily. Edwin and Roy Komarek (1938) took thirty-seven specimens in the Smoky Mountains. Here the mice were found even on the high grassy balds, in rocky outcrops on the summit.

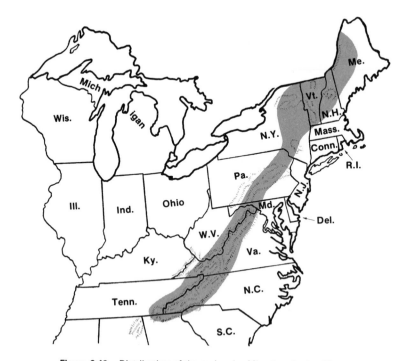

Figure 6.49. Distribution of the rock vole, *Microtus chrotorrhinus.*

The habits are but imperfectly known. These microtines live in shallow burrows and runways which thread about rocks, and in some places appear to be partial to a ferny habitat.

The young are born from early spring well into the fall, but seldom exceed three or four at a birth.

Locally these mice may be quite abundant, living in small colonies, but intensive trapping will soon take every member of the colony. They are absent from extensive areas which appear suitable for their maintenance.

Whitaker and Robert L. Martin examined the stomachs of forty-seven individuals from New Hampshire, New York, Labrador, and Quebec, and found the most important food items to be bunchberry, *Cornus canadensis,* making up nearly half the stomach contents; followed by unidentified green vegetation, lepidopterous larvae, and wavy-leaved thread moss, *Atrichum undulatum.* They will also feed on small rootstocks, green grasses, fresh shoots, and such berries as are acceptable. In Minnesota blueberry leaves and stems and Clinton's lily plants were heavily eaten.

The thrill of trapping a rare mouse or shrew is graphically told by Morris M. Green (1930). He writes:

Near Opperman's Pass, southwestern Wyoming County [Pennsylvania], in mid-October, 1927, my attention was drawn to a north-facing swale, filled with a tangle of ferns, moss-covered logs and boulders, shaded by yellow birches. The tinkling of an underground stream, beneath the boulders, could be plainly heard. A geologist might have said that this was a glacial moraine. There were many cozy little nooks, between the boulders, for little forest folk, so I carefully placed a dozen mouse traps there. The next day there were a New York deer mouse, a short tailed shrew, a red-backed mouse, in some of the traps. Peering down a cavity at one trap, there appeared to be a half-grown meadow mouse there. Drawing the trap out in the sunlight, my heart beat fast when I saw that the mouse had a saffron colored nose with a brown body. Good luck had enabled me to record the first specimen of the rock vole from Pennsylvania.

Prairie Vole. *Microtus ochrogaster* (Wagner)

DESCRIPTION.   The prairie vole is similar in appearance to the common eastern field mouse, *Microtus pennsylvanicus,* from which it may be distinguished by the presence of only five plantar tubercles and the crown of the third upper molar, which has but two irregular triangles between the anterior and posterior loops. Additional characters which are usually distinctive are the short tail, the grizzled appearance of the rather coarse dorsal pelage and the usually buffy belly (some individuals are white-bellied). The ears are small, nearly hidden in the fur. The feet are pale buffy, tail similarly colored except for a narrow dorsal stripe which is slightly darker. Measurements of ten adults from Illinois average: total length, 148 (134–162) mm; tail, 31 (25–37) mm; hind foot, 19 mm (17–22) mm; weight, 25–55 grams. Ohio individuals often have white underparts and have been described as *M. o. ohioensis*.

DISTRIBUTION. The prairie vole occurs from central Ohio and northern West Virginia westward through Wisconsin and northwestern Tennessee (Figure 6.50).

HABITS. The meadow mouse of the prairie country does not differ greatly in habits from the eastern species. It appears to prefer drier, more sandy situations, and has invaded much of Indiana and Illinois where farm-land has replaced the forest. During late June 1940, Hamilton camped near a small stream outside of the town of Gladstone, Illinois, a few miles from the Mississippi River. Here, on a little knoll surrounded by water, he trapped six of these mice in short order. They were feeding on the sedges and various grasses that grew to the water's edge.

   In Clark County, Wisconsin, F. J. W. Schmidt (1931) found these mice abundant in sandy plains and sandy slopes of sandstone mounds and in the woods of jackpine and jack oak; there was no heavy covering of grass in either of these habitats. The mice were living in small colonies on large knolls

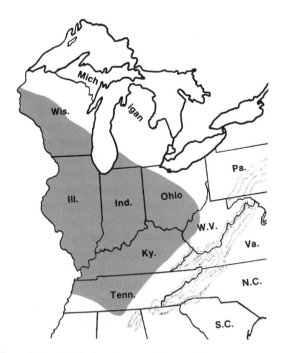

Figure 6.50. Distribution of the prairie vole, *Microtus ochrogaster*.

formed by the uprooting of trees. These colonies numbered three or four adults and several young, and were characterized by the great extent of the burrows. In Indiana, this species and *M. pennsylvanicus* share the grassy fields, with *ochrogaster* existing primarily in the drier, more sandy fields with smaller amounts but often more variable vegetation, and *pennsylvanicus* living in the wetter meadows with heavier ground cover, often of nearly pure grass. Especially in central Indiana, the two fairly often occur together where the habitats blend or interdigitate.

This species is said by Kennicott to build its winter nests in old anthills, or, if built directly on the uncultivated prairie, they are characterized by little mounds of earth at the entrance. The burrows are reasonably shallow but are remarkable for the numerous and complicated chambers and side passages of which they are composed. The bulky nest of dried grasses is placed in one of these chambers, and has but a single small opening on one side.

The reduced number of teats (six) in this species argues for smaller litters than in *M. pennsylvanicus,* and indeed this is the case. The mean number of embryos in eighty-five pregnant females was 3.5 (range of one to seven); but most had three (thirty-one individuals) or four (twenty-five individuals). Pregnant females were found in every month, but fewer pregnancies occurred in December to January and June to July. The evidence points to a breeding season extending from late March to October.

The feeding habits are not unlike those of the eastern *M. pennsylvanicus,* except, in keeping with greater numbers of plant species, *M. ochrogaster* has a more variable diet, although it still consists almost entirely of finely ground vegetation. Earl G. Zimmerman found bluegrass, clover, Lespedeza, old witchgrass (*Panicum*), fleabane, plantain, fescue, and black medic to be some of the more important foods at an Indiana locality.

These mice appear to store food for winter habitually, and several writers mention finding sizable quantities of tubers, roots, and small bulbs in the winter chambers. They move into the shocks of corn as winter approaches, feeding on the germ and destroying great quantities in this fashion.

Where they occur in orchards, the prairie voles often cause considerable loss among young fruit trees by girdling the base. Fortunately they are partly held in check by numerous enemies, which include snakes, hawks and owls, and several predatory mammals.

Pine Mouse. *Microtus pinetorum* (LeConte)

DESCRIPTION.   *Microtus pinetorum* is a small, robust, short-tailed mouse with very soft, short dense fur which is almost molelike. The eyes and external ear are much reduced, the latter usually hidden in the fur. The tail is only slightly longer than the hind foot. It has five plantar tubercles and four mammae; the skull is flat and wide, with quadrate braincase and short rostrum; the palatal slits are relatively short, ending well ahead of the tooth row. The skull of this species is very similar to that of *M. ochrogaster*. Measurements average: total length, 110–115 mm; tail, 17–19 mm; hind foot, 15–16 mm. Upperparts are bright brown or chestnut, with a distinct sheen on the fur, somewhat lighter on the sides; belly dusky to silvery gray, the bases of the hairs plumbeous; tail brownish above, pale below; feet pale gray (Figure 6.51). In the northern parts of its range the pine mouse is larger and duller than in the southeast. It is dull chestnut above, paler on the sides; underparts silvery to slate gray. Measurements of fifteen adults from Ithaca, New York, average: total length, 126 mm; tail, 19 mm; hind foot, 18 mm; weight, 25–35 grams. New Jersey specimens are smaller. Individuals from the southeastern part of our area are small, characterized by the relatively large ears which stand above the fur; colors dark and rich, and the fur short and dense. In eastern Kentucky and Tennessee, this species is darker and richer in color than the surrounding populations.

In southwestern Wisconsin individuals are of large size, have large ears, relatively long coarse fur, and dull colors, and are not so dark as in other northern populations.

DISTRIBUTION.   The pine mouse occurs from New Hampshire to Georgia and northern Florida, westward to Wisconsin and Mississippi. It has a wide choice of habitat, from sea level in the south to the spruce and birch of the higher northern mountains (Figure 6.52).

HABITS.   Although the pine mouse is basically a woodland form, few small rodents are more adaptable than it is. About coastal bays, it occurs to the edge of tidewater. It occurs in great numbers in the dry fields and truck

Figure 6.51.   Pine mouse, *Microtus pinetorum.*

Figure 6.52. Distribution of the pine mouse, *Microtus pinetorum.*

gardens of southeastern United States, while farther north it may occupy a diverse habitat, actually invading the cool forests of yellow birch and hemlock in the heights of the Berkshires in western Massachusetts. It is most abundant in woods in the sparsely wooded regions of the middle states, where the soil is loose and friable. Its name is a misnomer, for the species is seldom found in stands of pines. It can be most easily trapped by sinking mousetraps crosswise of the burrows with the treadle at the level of the floor of the burrow. We usually poke the fingers into loose, moist woodland soil of leaf mold to find burrows. Best results are usually obtained when the holes containing the traps are covered with bark or other material. The mouse often occurs where there is a thin mat of herbaceous vegetation, but also where there is little.

Ray T. Jackson indicates that in northern Florida this mouse usually selects areas covered with dense trees and shrubs, mainly sand pine and scrub oak. No runways were found in openings with no trees. Burrows ranged from about three-fourths of an inch to an inch in diameter and wove among the roots of the trees and shrubs. One nest under an old board was of palmetto bark, and lined with silky, fibrous material, much like the inside of a milkweed pod. Around the nest were hulls of about two hundred pine seeds.

The short ears, sleek molelike fur and strong feet are all adaptations for a fossorial life, and *M. pinetorum* has need of them, for it seldom ventures any distance above the ground, and then only to chase over the shallow surface runways into another burrow.

In digging in the forest floor or dry fields, the pine mouse threads its way just beneath the thick carpet of leaves, the latter forming a thin but substantial ceiling to the burrow. The mouse may on rare occasions tunnel to a depth of a foot or more, but most of the tunnels do not exceed three or four inches in depth.

The nests are made of dead grasses and leaves, and are placed beneath a log or more often in a snug chamber several inches below the surface. In these the one to four young (seldom more) are born; the breeding season lasts from early March well into November. Occasionally the mice breed throughout the winter, even in the northernmost part of their range. The pine mouse has four teats, half the number of the meadow vole. Although its rate of increase is considerably less than that of the meadow vole, it probably has fewer enemies, for the subterranean habits secure it in a measure from hawks and owls. The young, as those of most other mice, are blind, naked, and helpless at birth, but grow rapidly, the eyes opening on the ninth day. A few days later they are weaned. If the nursing female is frightened from the nest, the young cling tenaciously to her teats and are dragged considerable distances without losing their grasp.

The food of the pine mouse includes succulent roots and tubers, green leaves and stems, peanuts, potatoes, and other truck crops. These mice are fond of any fleshy roots and practice a limited storage. We have found them partial to the blanched roots of many common grasses of the farmyard. They appear to relish even more the dead bodies of kindred species, and will make short work of their own kind if it be caught in a trap.

This little rodent is a serious pest of the orchardist. It girdles the roots of fruit trees, and this activity may be extensive enough to cause widespread damage in the fruit belts of eastern United States. Damage of this nature has been particularly severe in the Shenandoah Valley. We have seen uprooted apple trees wholly girdled, the smaller rootlets completely eaten or cut from the tree. On Long Island and elsewhere severe damage to potatoes has been caused by this species. It is a stubborn pest, and one rather difficult to control, but poison baits placed directly in the runways usually are effective in reducing the numbers.

Round-Tailed Muskrat. *Neofiber alleni* True

DESCRIPTION.   In general appearance like a small muskrat, *Neofiber* has a dense waterproof coat, small ears almost hidden in the fur, a scaly *round* naked tail almost devoid of hair, and slightly webbed hind feet. Upperparts are generally dark uniform brown, the underparts varying from almost white to buffy. The darkest individuals occur in the southern part of peninsular Florida.

The average measurements of four Florida adults are: total length, 319 mm; tail, 123 mm; hind foot, 44 mm. Adults weigh about 300 grams. (Figure 6.53).

DISTRIBUTION.   *Neofiber* occurs from southeastern Georgia south to the Florida Everglades (Figure 6.54).

HABITS.   The round-tailed muskrat replaces the muskrat in the southeast. Here in the sphagnous bogs of southeastern Georgia and the coastal marshes

Figure 6.53.   Round-tailed muskrat, *Neofiber alleni*.

Figure 6.54. Distribution of the round-tailed muskrat, *Neofiber alleni.*

and weedy shores of Florida lakes, it makes its home, occurring in some numbers in favored localities.

Among the salt savannas bordering the inflowing streams along the Indian River of Florida, *Neofiber* is abundant. Frank M. Chapman found, among the berry fringe of red and black mangrove, tangled masses of grass from two to three feet high and densely matted underfoot. In such places, *Neofiber* constructs large woven nests, placed in hollow stumps or trunks of the mangroves, or less often in the open savanna. These nests are sometimes as large as a basketball. Dale E. Birkenholz (1963) has made extensive studies of this species in Florida. He found houses to range from seven to twenty-four inches in diameter, averaging about twelve inches, with an internal chamber about four inches in diameter. A house constructed for normal living had walls two to three inches thick, with floor level at or up to about two inches above water level. The cavity was unlined and moist. Females about to give birth altered this type of house, making the walls much thicker, raising the floor, and lining the chamber with dry, fine-textured grasses. Sometimes still more complex houses are built. Generally two openings lead, on either side, from the single chamber to tortuous galleries below the matted vegetation. In the freshwater ponds near Gainesville, Florida, large nests are built in the water bushes, *Cephalanthus,* or rest on the surface in water one or two feet deep. Unused houses are often used by rice rats and cotton rats. In the great Okefinokee Swamp of southeastern Georgia, *Neofiber* is largely restricted to the so-called prairies—level, almost treeless bogs or marshes. For the most part these are covered with water and aquatic plants rooted in the muck and peat which has accumulated to a depth of several feet above the sand bottom.

In some parts of Florida these rats tunnel in both cultivated and neglected fields, in cane patches and even in dooryards and gardens. A. H. Howell found the burrows ramifying through the friable peat in all directions but apparently extended to no great depth. Sometimes, perhaps because of low water, these animals live in burrows rather than houses.

Neofiber is much less aquatic than the muskrat, but is an excellent swimmer and does take to the water readily, swimming and diving with ease. The tail is said to gyrate in a peculiar manner, the tip describing circles.

A feeding platform is constructed in shallow water, where succulent grasses may be pulled within reach. The discarded plant is added to the platform so that the feeding station eventually obtains considerable bulk. Harper describes the feeding platforms of Georgia animals as being slight, smooth-worn mounds of sphagnum, peat, herb stems, and the like, with two tunnels leading downward into the bog on opposite sides. Here the round-tailed muskrat brings crayfish, seedpods of *Iris* and *Sagittaria,* and succulent root stocks.

Neofiber breeds throughout the year; there are no well-marked breeding cycles, although there is increased reproduction when habitat conditions are good. Gestation is about twenty-six to twenty-nine days. The number of young per litter in forty-eight pregnant females and fifty-six females with placental scars examined by Birkenholz ranged from one to four, with two being the most common. Two litters may be produced in a three-month period, and it is possible that some females might produce four or five litters per year.

Water moccasins are an important predator on this species, and marsh hawks take their toll of *Neofiber,* along with barn owls.

Birkenholz found from observation of the contents of 330 stomachs and from food remains at feeding sites that this species has a rather restricted diet. Maidencane was the principal food species in his study sites, but occasionally *Sagittaria lancifolia, Hydrotridia caroliniana,* and *Brasenia schreberi* were eaten. *Brasenia* might have been of considerable importance if it had been more abundant. Elsewhere *Nymphaea, Pontederia, Mariscus, Sporobolus, Panicum, Peltandra,* and other plants have been recorded. Crayfish remains are often found on feeding platforms, but have never been found in stomachs. They are likely left by rice rats, which often use the platforms.

Muskrat. *Ondatra zibethicus* (Linnaeus)

DESCRIPTION.   The muskrat is a large, robust rodent with short legs, large hind feet, the toes partly webbed; the tail is long and compressed laterally, scaly and sparsely haired. Ears are small, almost hidden in fur; pelage dense, the underfur soft and overlaid by long guard hairs, the entire pelage waterproof. General color is rich brown above; paler below. Back and sides are rich brown, the back darker owing to the black-tipped guard hairs which arise above the shorter underfur. Underparts are pale, shading to white on the throat, but quite variable. There is a blackish chin patch. In late winter the guard hairs, particularly along the sides of the body and flank, are tipped with gold. Average measurements of ten adults from central New York are: total length, 546 mm; tail, 243 mm; hind foot, 77 mm; weight, 1.7–3 pounds (780–1360 g). Muskrats from the extensive cattail marshes are much larger and heavier than those from pasture streams (Figure 6.55).

The muskrat of the Louisiana coastal marshes is slightly smaller than typical *zibethicus*; colors are duller, lacking the reddish tints of more northern forms. Average measurements of ten adults are: total length, 547 mm; tail, 233 mm; hind foot, 78 mm; weight, 1.5–2 pounds (700–900 g).

DISTRIBUTION.   The muskrat is found from Maine and Michigan south to southern Louisiana, central Georgia, northern South Carolina, and Virginia (Figure 6.56).

HABITS.   The amphibious muskrat occurs wherever water and food plants provide satisfactory conditions of life. It is most abundant in extensive marshes, where reasonably shallow water supports a varied assortment of cattails and other food plants. Streams through pastures and wooded swamps also provide a homesite, but in such places the rats are less abundant. The extensive marshes of Delaware and Maryland are favorite resorts of these animals, and on the Gulf Coast, in southern Louisiana, untold numbers live in the extensive marshes.

The muskrat is seldom observed far from water. Its broad webbed hind feet and flattened tail act as efficient oars and scull; the muskrat may remain submerged for several minutes.

Figure 6.55.   Muskrat, *Ondatra zibethica.*

Figure 6.56. Distribution of the muskrat, *Ondatra zibethica*.

In marshes and swamps, the muskrat builds a house of cattails, sedges, water plants, and the thick mat of decayed vegetation which lies below the water level. One in an Indiana pond contained much duckweed. These houses are usually constructed in water not more than two feet deep and are built up merely by the accumulations of such trash. The lodge may be built in a rather shallow part of the swamp or actually amidst a clump of willows or other shrubs. The removal of surrounding vegetation by the muskrat results in a deepening of the channels. A snug dry chamber, slightly above the water level, serves as a sleeping chamber, and from this one or two plunge holes lead to the water. These lodges vary greatly in size and shape; usually they are dome-shaped structures, but houses in Maryland often have slanting or flat tops. The houses harbor from one to ten or more individuals. Smaller feeding lodges are constructed, large enough to house only a single rat at a time. Here food is brought and eaten in comparative safety.

Wherever conditions are unsatisfactory for the construction of a house, a tunnel is dug in the bank below the water surface, leading to an enlarged chamber well above high water level, where a warm nest of grasses is constructed. Another entrance may open on the bank.

The muskrat is a prolific creature. In the northern states the first litter is born in late April or early May, after a twenty-nine- to thirty-day gestation. Breeding occurs throughout the year in Louisiana, although the heaviest breeding occurs from November to April. The young rats are born with a fine covering of hair. Growth is fairly rapid, the young reaching a good size and shifting for themselves when little over a month old. Several litters are produced in a season, the young numbering from one to nine. Three or four young is the average litter in the Gulf states, although five or six is more usual in the north.

Ted O'Neil (1949) found that in the southern Louisiana marshes, muskrat houses with pregnant females or newborn can usually be recognized because they are freshly worked with plastered muddy peat. Females copulate about ten days after giving birth, and a new nest is constructed in the same house. This process is repeated until three or four nests are present; then these nests are used over and over. The adults and young all work together on house maintenance. As young reach sexual maturity, they are forcibly evicted. Thus the colony may expand, or new colonies may be started. Usually a house contains the adults and two to four young, but sometimes a second house is connected to the first, forming a "double house," and ten to fifteen muskrats may live and work together.

During the breeding season the prepucial, or musk, glands enlarge and secrete a very pronounced, rather pleasant, odor. The function of these large glands is not well understood but it is thought that the scent draws the sexes together.

Muskrats are exemplary feeders. Chiefly vegetarians, they eat the roots and blanched stalks of cattails, three-square grass, sagittarias, and the leaves, stems, and fleshy roots of many aquatics; they even invade neighboring fields to cut down herbaceous plants or tear down the growing corn. The most important foods on the extensive gulf marshes are three-squares (*Scirpus*), needle grasses (*Juncus*), and paille-fin grasses (*Panicum* and *Spartina*). The large fresh-water clams are relished, and muskrats eat fish and crustacea, but these usually form a minor share of the food.

The muskrat is plagued by many predators. Marsh hawks coursing low over the marshes capture many luckless young rats; the larger owls find good hunting in the swamps. Foxes range over the frozen sloughs during the winter or course the higher levels in the spring and summer, intent on a feast of rats. The mink is the most dreaded enemy, for it enters the houses and bank tunnels. This predator is usually more than a match for the largest rat. Water moccasins destroy some muskrats, and the larger turtles and even alligators occasionally prey on this species.

A new problem for southern muskrats has been the introduction of the nutria which, of course, has habits similar to those of the muskrat, placing the two species in direct competition. In addition, the marshes have deteriorated. The net result is that the nutria has now replaced the muskrat as the most important fur-bearing animal in Louisiana.

Elsewhere, the muskrat is still the most important of our native fur bearers. Formerly worth only a few cents, the pelts skyrocketed to four dollars during the boom following World War I, and from 1933 through the 1940s the price averaged a dollar or more for prime pelts. Many trappers of Maryland's east shore, the coastal marshes of Louisiana, and even the extensive swamps of New York, Michigan, and Wisconsin make a partial living from the muskrat. In addition, the carcass is sold in some numbers in the markets of Wilmington, Baltimore, and Washington. Average prices of muskrat skins in Indiana were $.99, $1.46, $2.36, $1.89, $2.49, and $2.98, for the years 1970 through 1975 respectively. According to George Lowery (1974), more than a million and a half pelts were harvested in the 1968–69 season in Louisiana, the pelts averaging $1.10. A million pelts taken during the 1976 New York season averaged $5.20 per pelt.

Key to the Bog Lemmings, Genus Synaptomys

A. None of the hairs at base of ears appreciably brighter than remainder of pelage. Lower molars with triangles on outer sides; palate with broad, blunt median projection . . . .*Synaptomys cooperi*

AA. A few hairs at base of ears distinctly rust colored. Lower molars without triangles on outer sides; palate with sharp, pointed median projection . . . . . .*S. borealis*

Southern Bog Lemming. *Synaptomys cooperi* Baird

DESCRIPTION.   *Synaptomys cooperi* is a small, robust, short-legged vole with a large head and a very short tail, which is scarcely longer than the hind foot and sometimes actually shorter, and broad heavy upper incisors with a shallow groove on their outer anterior surface (Figure 6.57). The nail of the first (inner) digit of the forefoot is flat and strap-shaped. Pelage is rather long

Figure 6.57. Front view of upper incisors of *Synaptomys* showing shallow groove.

and shaggy; dorsal color mixed brown, gray and black, with a hint of dark yellow, lending a grizzled appearance, the whole tone brownish; sides and underparts silvery gray, the bases of the hairs darker; tail brownish above, whitish below but not sharply bicolor. Fifteen adults from eastern New York and Pointe au Baril, Ontario, average: total length, 121 (114–130) mm; tail, 16 (13–18) mm; hind foot, 17.5 (17–18) mm. Adults weigh 24–35 grams. Average measurements of nine adults from the Great Smoky Mountains are: total length, 127.2 (120–136) mm; 24 (20–27) mm; hind foot, 20.4 (19–21) mm; weight, 26–36 grams (Figure 6.58).

DISTRIBUTION.   The southern bog lemming occurs in the eastern United States, from Maine to the Dismal Swamp in North Carolina and the Great

Figure 6.58.   Southern bog lemming, *Synaptomys cooperi.*

Figure 6.59. Distribution of the southern bog lemming, *Synaptomys cooperi.*

Smoky Mountains, westward through southern Illinois and Wisconsin (Figure 6.59).

HABITS. The little bob-tailed lemmings have an extensive range in boreal America, descending in the eastern mountains as far south as the towering mist-shrouded Smokies. Dry hillsides with a growth of bluegrass, fields matted with a canopy of weeds, grassy areas with interspersed brush and small trees, the dense woods of hemlock and beech, all harbor these little mice.

In the fields of blue grass runways are constructed above ground, crisscrossing one another; the grass is cut and trimmed so that the highways are kept smooth. On the northeastern forest floor the bog lemmings tunnel just beneath the leaf mold, pushing beneath the black soil or running through the hidden tunnels of the big hairy-tailed mole. In New York these little animals can be caught by sinking mousetraps crosswise of underground burrows in hardwood forests. We have never taken any in Indiana in that situation; in fact, in Indiana they are almost never taken in woods and are usually in bluegrass, little bluestem, or in other grassy stands. One of the habitats in which this species is least likely to be found is in bogs.

Wherever they occur, bog lemmings are found in company with other small mammals—red-backed mice, deer mice, prairie voles, and various shrews and moles may be part of the community. *Synaptomys* often occupies the same burrows as these others. Nests are constructed of dead grasses and leaves, often well hidden beneath the surface at a depth of several inches, or less often directly on the ground wherever there is sufficient cover to conceal them. Occasionally the nests are lined with fur.

Bog lemmings are sociable little beasts and usually occur in colonies the populations of which range from a few to several dozen. Of a sizable area offering similar habitat, a rather small part will be occupied by a little colony and the remainder will be devoid of these mice.

Perhaps the most prominent "sign" of these mice is the little piles of grasses or sedges which are found in the burrows or runways. These cuttings are of match length, and are found in varying degrees of freshness. The little oval bright green droppings are scattered throughout the burrows or are massed at certain points of the runway.

The breeding habits are similar to those of other small voles. In the mountains of North Carolina they are known to breed from February to November; farther north the season of reproduction is equally long, half-grown mice being secured at all seasons of the year. The number of young usually varies from one to four; thus the litters are on the average somewhat smaller than those of allied species.

According to LeRoy Stegeman (1930) the young are blind and naked at birth, and weigh about 2.5 grams. Hair appears on the sixth day and the eyes open on the twelfth day.

The food consists largely of plant material. All the stomachs we examined contained a finely chewed mass of green vegetation; in addition the seeds of raspberry and the mycelial threads of a fungus, *Endogone,* have likewise been found.

Synaptomys usually refuses all lures, but an unbaited trap carefully set across the runway will often prove successful in their capture. This vole is relatively rare in collections, but it is by no means an uncommon small animal, its colonial habits making it perhaps more restricted than other small species.

Owls, hawks, predatory mammals, and snakes all prey upon this mouse.

Northern Bog Lemming. *Synaptomys borealis* (Richardson)

The lemming mice of the *borealis* group have been placed in the subgenus *Mictomys.* They differ from *cooperi* in lacking closed triangles on the outer surface of the mandibular molars. Moreover, the incisors are much more

Figure 6.60. Distribution of the northern bog lemming, *Synaptomys borealis.*

slender than in *cooperi,* the maxillary incisors often having the outer corners unworn and prolonged into sharp splinters of enamel. Color above is dull brown, often with olive wash, brighter on rump, anteriorly more grizzled; tail bicolor. It is known from the type locality at Fabyans, at the base of Mount Washington, New Hampshire, and from northern Maine (Figure 6.60).

HABITS. The habits of *Synaptomys borealis* are not significantly different from those of *S. cooperi.*

MURIDAE

(Old World Rats
and Mice)

Various rats of the genus *Rattus,* and the house mouse, *Mus musculus,* have been introduced into the New World since the American Revolution, and these pests are now well established over most of the country. They are characterized by the long, naked tail, the typical mouse- or ratlike form, and three molar teeth on each side of both jaws. The upper molars differ from those of native rats and mice in having three longitudinal rows of tubercles (Figure 6.61), a character which can be observed even in well-worn teeth.

Key to Rats of the Genus Rattus

A. Tail longer than head and body; first upper molar with distinct outer notches on first row of cusps . . . .*Rattus rattus*
AA. Tail not longer than head and body; first upper molars without such notches . . . .*R. norvegicus*

Black Rat. *Rattus rattus* (Linnaeus)

DESCRIPTION. The black rat is similar in form to the Norway rat, but with shorter nose; color darker, grayish black above, sooty below; tail more than half total length. Measurements of twelve adults from Alabama, Florida,

Figure 6.61. Upper molars of Norway rat, *Rattus norvegicus,* with tubercles in three series (left) contrasted with those of the upper molars of the deer mouse, *Peromyscus leucopus,* which have tubercules in two series.

Georgia, Massachusetts, New Hampshire, and Virginia average: total length, 369 (327–430) mm; tail, 193 (160–220) mm; hind foot 35.5 (33–39) mm. The weight of large males is about 200 grams (Figure 6.62).

The roof rat, *Rattus rattus alexandrinus,* is similar in body characteristics to the black rat, but the color is more like the Norway rat. The best character is the long slender tail which is considerably more than half the total length of head and body and the whitish belly, often washed with yellow. Measurements of eleven adults from Washington, D.C., Tennessee, Florida, Georgia, North Carolina, South Carolina, and Louisiana average: total length, 398 (354–435) mm; tail, 215 (191–238) mm; hind foot, 36 (32–39) mm. This race and the black rat exist side by side in the south, and apparently interbreed freely, for specimens often show characters of each.

DISTRIBUTION.   Generally distributed along the coast and established in many inland cities, black rats are more abundant in the southern states and have been recorded from most of the states east of the Mississippi. Colonies have been reported at Springfield, Massachusetts, and many cities on the Great Lakes. However, the Norway rat appears to compete successfully with the black rat; when the former species becomes established, black rats often disappear.

HABITS.   The black rat often falls prey to its large and more barbaric rival, the ferocious Norway rat. Being a more adept climber, it has managed to maintain itself in trees, the roofs and upper stories of buildings, and to a large extent in ships. Hamilton has found nests of the black rat in coconut palms on small islands off the coast of eastern Cuba. It and the roof rat are

Figure 6.62. Roof, black, and Norway rats.

said to nest in similar situations in southern Florida. The roof rat is most abundant in the deep south; it lives chiefly about the roofs of dwellings and in smokehouses and outhouses, occurring less commonly in the open fields.

The black and roof rat apparently are somewhat less prolific than the Norway rat, producing fewer and smaller litters. Nevertheless, their rate of increase is sufficient to make them a pest of major importance and their archenemy, man, is forever attempting to reduce their numbers.

The black rat is a menace of first importance in harboring the infectious agents of plague and typhus fever. Wherever it occurs, public health officers must forever be on the alert, so that a pandemic of these dread diseases will not take a frightful toll of human beings. Epidemic typhus fever of the Old World is transmitted by the rat-borne louse to man. New World endemic typhus fever, although milder, is nevertheless frequently fatal. It is transmitted from rats to man by the rat fleas, *Xenopsylla cheopis* and *Nosopsyllus fasciatus*. During the late summer and fall of 1942, army duties took Hamilton to Texas, where he was assigned to work on the control of typhus. By mid-October, more than nine hundred cases of typhus fever had been reported in Texas, and the disease was widespread in Georgia. A severe localized epidemic occurred in August, 1941, in Lavaca County, Texas. More than one hundred cases were reported in a single week, and several of them resulted in death. Fortunately, with the spread of the aggressive Norway rat, the black rat and roof rat appear to have diminished in numbers. Perhaps the establishment of the Norway rat, particularly in the South, will prove to be a blessing in disguise!

In the eastern United States, there are three subspecies, and in Louisiana, at least, all three occur, sometimes even at one locality, and intergrades do occur. However, presumably each of the three subspecies has been introduced from different localities where they are still found and are separated (perhaps) by primary isolating mechanisms. The occurrence of three separate subspecies together through recent introductions such as this should give some enterprising mammalogist in Louisiana a chance to study an interesting evolutionary situation.

Norway Rat. *Rattus norvegicus* (Berkenhout)

DESCRIPTION.   The Norway rat is a coarse-furred rat, with prominent naked ears and nearly naked scaly tail, which is not longer than head and body. Molars of the upper jaw have tubercles in three longitudinal rows, as in the house mouse. General color above is brown, with scattered black hairs, darkest on the middle of the back; underparts pale gray or grayish brown. The Norway rat might be confused with the rice rat (*Oryzomys*), as both sometimes occur in the same habitat. The latter has softer fur, lighter underparts, and a much slenderer tail; it also lacks the characteristic Roman nose of *Rattus*. Measurements of fifty adults from New York and Washington, D.C., average: total length, 399 (320–480) mm; tail, 187 (153–218) mm; hind foot, 41 (37–44) mm; weight, 300–540 grams (Figure 6.62).

DISTRIBUTION. The Norway rat is cosmopolitan; it was introduced into the United States in the latter part of the eighteenth century and is now established in every state in the Union.

HABITS. The rat is the greatest mammal pest of mankind. It has caused more deaths than all the wars of history. It harbors lice and fleas, dread disseminators of the plague, typhus, trichina, infectious jaundice, and many other but scarcely less serious diseases. These animals are usually a contributing factor of first importance in the spread of pandemics during war.

Rats occur wherever there is an abundance of food and shelter. Without these prime requisites, they soon move on to more favorable regions. Rats occur in the subways and crowded tenements of the great metropolitan districts and in the corn and grain fields of the farm country. They are frequently seen in the salt marshes of the Atlantic Coast, where edible flotsam is washed on the beaches.

Rats are largely nocturnal, leaving the shelter of their nests as dusk approaches. They are wary creatures, often occurring in incredible numbers on a farm, but such great numbers are seldom suspected, as they are adept at hiding or scurrying away at the approach of man.

The appetite of the rat is prodigious. It will eat a third of its weight in twenty-four hours, and often waste as much more if feeding in the poultry house. Few items are shunned. Its sharp teeth puncture the tin covers of jellies, and it consumes soap, candy, milk, meat, vegetables, poultry and eggs, and all grains; even cherries growing many feet from the ground are not exempt from the attacks of these ubiquitous rodents.

Some of the more important foods of 115 Norway rats from Indiana farms and granaries were grain seeds, mostly wheat (39.7 percent of the total volume), corn (20.2 percent), flesh (6.3 percent), green vegetation (5.3 percent), mast (5.0 percent), clover flowers (3.9 percent), and garbage (3.2 percent). It was estimated that rats eat perhaps 7,000 tons of cultivated grains per year from Indiana's granaries and larger farms.

Much has been written on the reproductive potentialities of the rat, a great deal of which has been gross exaggeration. Nevertheless, rats are among the most prolific of all mammals. If food is abundant, and shelter adequate, rats will breed throughout the year, although fewer litters are produced in the winter. The gestation period varies from twenty-one to twenty-three days, but may be prolonged when a pregnant female is nursing an earlier litter. The number of young in a litter varies considerably. Seven is the usual number, but we have counted from two to fourteen embryos in rats. Blind, naked, and helpless at birth, they grow rapidly, and the eyes open in fourteen to seventeen days. The young are weaned when three weeks old. The young rats commence to breed when three months old, although there is a record of an eight-week-old female giving birth to eleven young, all of which were successfully raised. They continue to breed until they have reached an age of one and one-half to two years, with each female producing an average of perhaps five litters per year. The life span of a rat may reach three years, which is comparable to ninety years in humans.

Man has waged warfare against the rat for centuries, but this pest continues to be an unmitigated nuisance, causing damage of many hundreds of millions of dollars each year. The many enemies, such as cats, dogs, snakes, hawks, owls, and weasels, all serve to lessen the rat numbers, but can not begin to keep populations down when environmental conditions are good for rats.

The rat menace should cause national concern. Every plagued community should organize an annual campaign to make residents rat-conscious, and to eliminate rat habitat. In this manner perhaps we can eliminate enough to reduce materially the losses they cause. As with *Mus musculus,* perhaps the one redeeming feature of this species is that it has given us the albino laboratory rat, widely used for medical and other research purposes.

House Mouse. *Mus musculus* Linnaeus

DESCRIPTION.   This cosmopolitan species is sometimes confused with our native species, as it is often caught in the fields and marshes. It can be identified by the amateur with certainty only by observing the crowns of the molars in the upper jaw, which have the tubercles arranged in three longitudinal rows. Its general color is grayish brown to brown, shading to lighter brown on the belly, which is often buffy. There is no line of demarcation between the dorsal and ventral coloration. The prominent ears and long, nearly naked and scaly tail are other distinguishing characters. Measurements of thirty adults from central New York and Washington, D.C., average: total length, 163 (148–205) mm; tail, 77.9 (69–85) mm; hind foot, 18 (16–20) mm; weight, 14–24 grams (Figure 6.63).

DISTRIBUTION.   It is cosmopolitan, found in all localities settled by man, and not infrequently well removed from human habitations. It was introduced from Europe probably about the time of the American Revolution, and is now widely distributed. This species is called the house mouse, and indeed, it often makes its residence in houses, barns, storage bins, and other buildings, but it should not be thought of as a species only or primarily of those habitats. It reaches its greatest abundance in the fields of corn, wheat, sorghum, and to

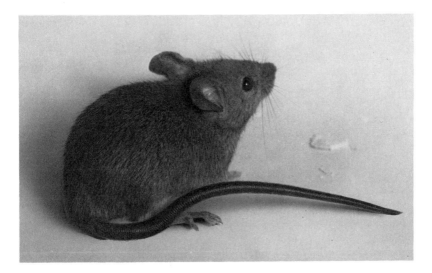

Figure 6.63. House mouse, *Mus musculus.*

a lesser extent, soybeans of the midwest. It is probably the most abundant small mammal of Indiana, because of the preponderance of such habitats. This mouse often takes up residence in grassy fields and waste lands, and the inexperienced collector may fail to recognize it under such conditions. It colonizes the cabbage palm thickets of Florida's west coast. It may be confused with certain forms of the harvest mouse, but the absence of grooved incisors will prevent mistaken indentification.

HABITS.   The house mouse exists in vast numbers in the habitats listed above, but only when good ground cover is present. It is, to a great degree, a nomadic species when living outdoors. When the ground is plowed or the crop is harvested the house mouse moves, perhaps to another field yet to be cultivated, or to a grassy or weedy field. It almost never enters the woods. House mice often enter midwestern homes particularly during fall harvesting time. They turn up in Whitaker's Indiana home every fall within a few days of harvesting of the large corn or sorghum field behind the back yard.

The persistent gnawing in the partition or the tell-tale little black pellets on the pantry shelf are often the only clue to the presence of this little beast until a few traps are set. Although a single individual will not pilfer any great quantity of food in one night, the combined depredations of several dozen over a few months' time may be considerable.

Unfortunately these mice, like their larger cousin, the Norway rat, are very prolific. It is not unusual for them to produce a litter of ten or eleven, although the usual number is but five to seven. The average number of embryos among forty-eight pregnant females from Vigo County, Indiana, was 6.02 (range of three to ten). Breeding continues without interruption from early spring until late fall, but is much curtailed or stopped altogether during the colder months. No pregnant females were taken in December or January in Indiana. After a gestation period of twenty-one days (somewhat longer if a pregnant individual be nursing a litter) the blind, naked, and helpless young are born. These grow rapidly and are fully weaned in three weeks. When less than two months of age these youngsters are ready to assume family duties.

In the fields, house mice eat many different foods, governed, of course, by availability. Some of the most important foods eaten by 458 house mice from Indiana were *Setaria* seeds, lepidopterous larvae, corn, wheat seeds, various other grass and weed seeds, and beetle larvae. In one field they fed heavily on the subterranean fungus, *Endogone*.

In the home, it would be easier to list the things these little pests do not eat than those which they do. Various grains and cereals, candy, paste from the bindings of books, meat, and almost anything edible left by the householder is taken. The damage is increased by their filth, which covers much that is otherwise spared by these mice.

House mice have various squeaks and calls, and even songs. Ernest Thompson Seton (1909) writes thus: "Out of the black darkness of a cupboard at midnight came a prolonged squeaking, trilling and churring, suggestive of a canary's song but of thinner and weaker quality. There could be no question that it was a 'singing mouse.' Many cases are on record." One night while checking traps in a weedy field at Willow Slough Fish and Game Area in Newton County, Indiana, we heard such a song over and over. We could not find the singer, but believe it was a house mouse.

Perhaps the successful establishment of this species over the entire world is due as much to its fortitude and endurance as to its adaptability. We have often caught these mice in the salt marsh, cattail swamp, or weed-grown field, and often they have been alive after hours in the trap. They are often caught by the tail, a foot, or loose part of the skin, and endure this captivity through the long cold night when a deer mouse or field mouse would quickly succumb. Whitaker on two occasions has caught two together in the same snap-trap.

Albino house mice are widely used in laboratories, where they make ideal subjects for biological, genetic, or medical research. This perhaps is their only redeeming feature.

ZAPODIDAE

(Jumping Mice)

Representatives of this family are found in the northern portions of both the New and Old World. They are characterized by the very large infraorbital

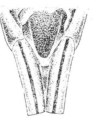

Figure 6.64. Upper incisors of *Napae-ozapus* showing deep grooves.

foramen, which allows transmission of muscles as well as the usual nerves, and in one of the two genera, *Zapus*, by the presence of four upper check teeth. The incisors are compressed and very deeply grooved (Figure 6.64). In general these animals are mouselike in form, but distinguished by their tremendously long tail and long hind legs which are adapted for leaping.

Jumping mice occur in forests, swamps and meadows; often they are common. They are profound hibernators.

Key to Genera of Jumping Mice

A. Upper cheek teeth four, the first very small; tail tip not white . . . .*Zapus*
AA. Upper cheek teeth three, tail tip white . . . .*Napaeozapus*

Meadow Jumping Mouse. *Zapus hudsonius* (Zimmermann)

DESCRIPTION. *Zapus* is mouselike in form, and the body is elongated posteriorly with long hind legs and very long tail, the middle toe of the hind foot being the longest. Pelage is rather coarse; yellowish brown above, with a prominent darker dorsal band, interspersed with many black-tipped hairs, the bases of the hairs slate colored; sides yellowish, with few black-tipped hairs, the hair bases white; underparts and feet white; tail brownish above, white below, sparsely haired and with a scanty tail tuft *which is not white*. *Zapus* has a very small upper premolar. Measurements of thirty adult specimens from New York average: total length, 213 (207–222) mm; tail, 128 (119–136) mm; hind foot, 31 (28–32) mm; May–June weights, 14–17 grams; September-October weights, 17–26 grams (Figure 6.65). The species is somewhat smaller in the southern parts of its range. Measurements of seven adults from Raleigh, North Carolina, average: total length, 191.3 mm; tail, 115.4 mm; hind foot, 28.3 mm.

DISTRIBUTION. The meadow jumping mouse occurs from Hudson Bay to South Carolina and Alabama, west to Iowa and Missouri (Figure 6.66).

HABITS. As the mower clatters through the still June meadows, little balls of fur, propelled by incredibly long legs, occasionally burst from its

Figure 6.65. Meadow jumping mouse, *Zapus hudsonius.*

path, or skulk off among the timothy stalks, for the dense canopy of waving grasses is the home of the meadow jumping mouse. Less frequently it inhabits the dry fields well removed from the watercourses, or small colonies may occupy the little grass- and brier-grown slashes in extensive forested regions. Where both *Zapus* and *Napaeozapus* occur, the meadow jumping mouse remains mostly in open areas, while *Napaeozapus* occupies the woods. However, in areas where *Napaeozapus* is absent, the meadow jumping mouse often becomes abundant in wooded areas, especially in thick herbaceous vegetation. It is often particularly abundant in patches of Touch-me-not, *Impatiens,* which often grow along streams or in wet places.

Under the cover of the broad-leaved vegetation these mice hide and feed during the day and night. They wander freely, seldom preparing or occupying runways, although in wet places ill-defined trails are apparently made by these mice. The well-constructed nest of grass and leaves is placed in an underground chamber, underneath a log, or less often in a clump of shrubs or stout weeds a few inches above the ground level.

This mouse is credited with making leaps of eight or ten feet, the powerful hind limbs acting as the propulsive force while the long tail provides balance. Such leaps are not the usual mode of progression, for usually the mice skulk among the vegetation or progress by a series of short jumps not exceeding a foot or so. In fact, we believe the long leaps are of the woodland jumping mice, not *Zapus.* On an open paved parking lot we were unable to stimulate meadow jumping mice to jump further than about three feet.

Jumping mice do not store food, but they are nonetheless active in garnering food in another manner. During September and October the larger animals acquire a substantial layer of fat (six grams or so). An individual takes about two weeks to put on this fat, and then enters the hibernation den. Thus, at any one time, surprisingly few of the animals taken in traps in fall have obtained their full fat complement. By October 20 to 30 essentially all of the animals have disappeared, with those yet too immature to put on an adequate fat layer apparently perishing during the winter season.

The winter quarters consist of a dry but often flimsy structure a foot or more below the surface, often in a bank, mound, or other raised area. In this

Figure 6.66. Distribution of the meadow jumping mouse, *Zapus hudsonius.*

retreat, the mouse curls in a tight ball, its head tucked under the body. The tail is curled about the body. In this rigid posture, the dormant animal passes the long winter, its temperature only a few degrees above freezing, its respiration slowed, and its heartbeat reduced to a few strokes each minute. Sometime in April, the time depending on the latitude, the mouse emerges from hibernation, but on cold nights it is inactive. It apparently hibernates throughout its range, but the period of dormancy is somewhat less in the southern states.

There are three peaks of breeding in this species in the eastern United States. Mating occurs soon after emergence in spring, and the three to six young are born in May or June, after a gestation of about eighteen days. Some individuals, probably mostly those not having a spring brood, produce a litter in July; many of the mice mate again in late summer with a litter being produced in August or even early September. We have found nests of young in open fields, the only cover being a weathered plank. It need not be so prolific as the other small mammals that are active throughout the winter and, as a consequence, are more susceptible to winter predation.

When meadow jumping mice emerge from hibernation in spring a variety of foods is eaten. Animal materials, especially beetles and cutworms, constitute about half the food, and various seeds about one-fifth. Few fungi are eaten. As the season progresses, more seeds, more fungi, and less animal materials are consumed. The seeds of various species of plants are the basic or staple food of this mouse, with seeds of various grasses, elm, Touch-me-not (*Impatiens*), chickweed (*Cerastium*), sheep sorrel (*Rumex acetosella*), Cinquefoil (*Potentilla*), and wood sorrel (*Oxalis*) often being important. The fungus *Endogone* and a few other tiny subterranean fungi are heavily utilized by this species, it forming 10 to 20 percent of the diet, especially in summer and fall. Jumping mice are particularly fond of berries and such fruits as grow in its meadow habitat, such as blueberries, blackberries, and strawberries. Often one comes on the little patches of inch-and-a-half or longer cut grasses, demonstrating the industry of this mouse as it cuts away sections of the close-standing timothy or a few other grasses to bring the seed heads within reach. After cutting the stalk off as high as they can, they pull the cut stems to the ground, again cutting until the head is reached. If one looks closely, the rachis and other parts of the seed head on the top of the pile can often be found.

The usual predators of mice take a sizable toll. Among the more important enemies are the raptorial birds, foxes, skunks, weasels, and snakes.

Woodland Jumping Mouse. *Napaeozapus insignis* (Miller)

DESCRIPTION.   This small, yellowish, mouselike rodent has a dark dorsal band, greatly developed hind legs, and a long, tapering, white-tipped tail. Color is yellowish-brown above, sprinkled with black hairs; the dorsal band dark brown interspersed with yellow-tipped hairs. Underparts are pure white; tail bicolor, dark brown above, creamy white to white below, the tail tip for varying length all white. The white tail tip, brighter colors, and the absence of a very small premolar in the upper jaw serve to distinguish this genus from *Zapus* (Figure 6.67). All jumping mice have prominently grooved incisors (Figure 6.64).

Figure 6.67. Woodland jumping mouse, *Napaeozapus insignis.*

Measurements of forty adults from New York are: total length, 227 (210–249) mm; tail, 139.5 (126–152) mm; hind foot, 30.2 (28–34) mm. Weights of May and June specimens average 20 grams, of September and October specimens just prior to hibernation 25 to 28 grams. South through the Alleghenies, it is smaller and considerably darker. Measurements of eight adults from the Smoky Mountains average: total length, 223 (185–233) mm; tail, 140 (120–148) mm; hind foot, 29.7 (29–30) mm.

DISTRIBUTION.   In eastern United States, this species occurs from Maine to North Carolina westward to Wisconsin. It is absent to the south except in the higher mountains (Figure 6.68).

HABITS.   Few small mammals can compare with the woodland jumping mouse in elegance of form or exquisite color. This brightly marked little rodent is seldom seen except in the trap, although it is by no means uncommon. Between us we have trapped several hundred along tumbling brooks, shaded by the tangled undergrowth and forest trees of central New York. This species prefers the forest, seldom if ever venturing into the haunts of its cousin, *Zapus,* although where there is an intermixing of meadow and forest herbaceous vegetation in open glades at the edge of a forest or other similar ecotone situation, the two may occur together. The woodland jumper skips nimbly over the clean shingle and sand bordering the tumbling little cascades of mountain slopes or lives in harmony with other little forest folk of the leaf mold. It makes no discernible runways, but sometimes utilizes the burrows of moles and the larger shrews, or seeks the shelter of rotting logs and fallen trees, whose exposed roots provide sanctuary. Several observers have seen this species make a jump of from ten to twelve feet, and we have seen them jump at least eight feet.

A nest is made of material at hand; it is usually placed several inches below the surface beneath a stump or slight eminence in the forest. The stomachs of 103 individuals of this species from New York contained tiny

Figure 6.68. Distribution of woodland jumping mouse, *Napaeozapus insignis.*

fungus of the genus *Endogone,* comprising about a third of the total food (Figure 6.69). This fungus is poorly known and difficult to find even by sieving soil samples, yet *Napaeozapus* and many other small mammals feed on it in quantity. We surmise they find it by their powers of olfaction. Seeds made up about a quarter of the food in stomachs, lepidopterous larvae and various kinds of fruit each about ten, and beetles 7.5 percent of the food. Seeds of Touch-me-not were very important when ripe, being easily identifiable by the brilliant turquoise color in the stomachs of animals eating this item. In addition, *Napaeozapus* samples the alder cones, the tiny buds, and mast, and searches industriously for tiny spiders and centipedes.

Feed well it must, for the first killing frosts signal the end of such bounty, and the fat creature retires to the depths, to pass the winter in a stiff ball, with the life processes reduced to a minimum. Perhaps they sleep in pairs, for the stomach of a November-caught skunk contained the remains of two of these mice. Usually half the year is spent in this dormant condition, for jumping mice are the most profound of eastern hibernators.

Three to six young are born in late June or early July, and some mice have

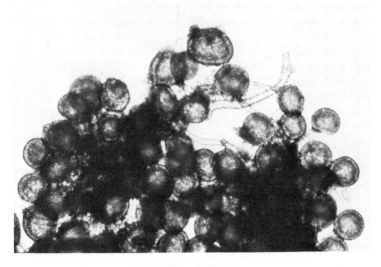

Figure 6.69. *Endogone.*

a second litter in August. Layne and Hamilton (1954) described the development of the young in this species. The dorsal color pattern is developed by the twenty-sixth day at which time the eyes open. The young are not weaned until a month old.

L. L. Snyder (1924) made some interesting observations on the home life of *Napaeozapus* while camping on Lake Nipigon, Ontario. He observed an adult mouse engaging in an eccentric dance which was not connected with the rut, for nest young of this individual were found a bit later. The entrance to the tunnel which harbored these young was closed during the day, at least during the time of nest-building and the rearing of the young. These mice are of little economic importance. They are seldom found in the haunts of man. Their usefulness lies in the fact that they probably reduce the numbers of forest insects and provide a partial source of food for predatory birds and mammals.

ERETHIZONTIDAE

(Porcupines)

Porcupines are characterized by their intricate folded and rooted cheek teeth, with flat crowns which have narrow ridges and wide reentrant spaces. The feet are well modified for an arboreal life. The large infraorbital foramen and stout skull, with heavy incisors, are also characteristic of the family (Figure 6.70).

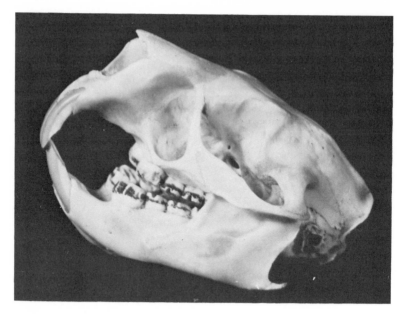

Figure 6.70. Skull of porcupine, *Erethizon dorsatum.*

The pelage is of three very different types: a woolly under fur, a coarser guard hair, and highly specialized spines or quills. These quills vary in size, but all are narrow and pointed, with undiscernible barbs at the free end. The short thick tail is heavily armed with quills.

Representatives of this family are found in North and South America.

Porcupine. *Erethizon dorsatum* (Linnaeus)

DESCRIPTION.   The porcupine, or, as it is often called, the hedgehog, is a thickset, heavy animal, with a small head, a short muscular tail clothed with spines above and stiff bristles beneath, and heavy feet, the soles of which are naked and rugose to aid in climbing. The pelage is brownish black above, some of the long hairs being tipped with white. These hairs are interspersed with numerous white or yellowish white, black-tipped quills which have minute barbs on the end. The shorter spines of the neck, rump, and tail are stiffer than those of the flanks and back. In winter the pelage is longer and darker, often obscuring all but the longest spines.

Average measurements of three adults from New York are: total length, 874 mm; tail, 147 mm; hind foot, 82 mm. Adults weigh from 10 to 20 pounds; very large males are said to attain a weight of 24 pounds or more (Figure 6.71).

DISTRIBUTION. The present distribution is now in the spruce and hemlock forests of northeastern United States from Maine to Pennsylvania, and in northern Michigan and Wisconsin. It may still occur in the wilder parts of West Virginia but probably never occurred so far south as the Smoky Mountains (Figure 6.72).

HABITS. The extensive forests of spruce, hemlock, and birch are the home of the blundering quill pig. It is freed from its winter fare of bark during the summer months and can then wander farther from the deep forest. But if a rock den or other suitable retreat is handy, a few acres will suffice, and this accounts for the local nature of porcupine distribution.

On the ground, its movements are slow and clumsy, nor is much speed made in the trees. It shows little concern for men, but if it is approached too quickly it shakes its tail angrily, and one must be wary, for the quills, although they cannot be thrown any distance, are shaken loose with ease, and they penetrate the skin, causing considerable pain.

While the porcupine is often abroad by day, it prefers the darkness of night. Its fondness for salt is proverbial; any article which has been handled by man is quickly chewed beyond repair. Hamilton recalls a night on the slopes of Mount Marcy in the Adirondacks when porcupines became a decided nuisance. His pack basket engaged their attention and he was busy through the early evening hours throwing sticks at the creatures in an effort to drive them

Figure 6.71. Porcupine, *Erethizon dorsatum.* Photo by Richard Fischer.

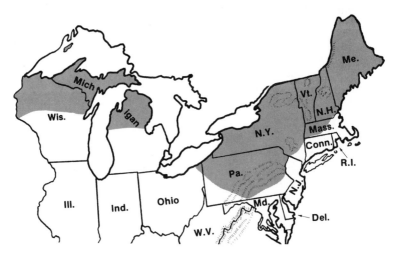

Figure 6.72. Distribution of the porcupine, *Erethizon dorsatum.*

off. In the morning, to his dismay, he found that the shoulder straps had been chewed from the pack. Deserted lumber camps are a frequent abode of these creatures; they are undoubtedly attracted by the greasy boards of the mess-room, which they soon chew to bits.

In northern New England a dozen or more are likely to frequent a rock fissure or abandoned quarry during the cold winter days, repairing to the nearby woods to feed during the night. Extensive damage may result from their girdling activities; white pine plantations are often killed by these big rodents. Zero weather holds no terror for the porcupine. Its dense coat is a sufficient guard against the severest winter storms, and only the protecting trunk of a large conifer guards it against the howling northern blizzard.

In winter the porky relies on the bark of many different trees. Hemlock and spruce seem to be preferred, but the bark of deciduous trees is not disdained. Hemlocks may be almost completely stripped of their smaller branches. Spruce are often completely girdled of a strip several inches in breadth. In the summer the porcupine's fare is more variable. Any log project-ing into a lonely lake of the forest wilderness is at some time visited by them, the better to secure the succulent lily stems and roots. It is said they follow the moose, feasting on the fleshy rhizomes which are torn from the mud by these great beasts. In the wilder agricultural lands, the porcupine visits the corn patch and alfalfa fields and may feed on the fallen apples. It is one of the very few rodents which may be considered a strict vegetarian.

Fall and early winter is the mating season; the mating act is not dissimilar to that of other mammals of comparable size. The gestation period is from 205 to 217 days; a copulation plug is formed. There is usually one young, rarely more. Young are born March through June. The young is remarkable for its precocious condition at birth; it weighs about a pound and is nearly twice the weight of a newborn black bear, or relatively forty times as heavy as a baby bear. Moreover, it is well clothed with fur. The quills are soft at birth, but harden within one hour. The eyes are open and some teeth have erupted. The mother giving birth is further protected by the young emerging head first and being in a sac.

The formidable armor of sharp spines would appear to provide substantial insurance against all enemies but man and the fisher. However, the red fox, bobcat, and perhaps other enemies manage upon occasion to kill the porcupine without suffering material harm, and the horned owl is known to tackle it when other foods fail.

The porcupine undoubtedly causes local damage, and in the white pine stands of Minnesota it may kill more than one percent of the annual growth. Under such conditions its reduction is justifiable, but for the most part its depredations are of little consequence. We have found the flesh to be tender and quite palatable.

CAPROMYIDAE

(Nutrias)

Hutias and nutrias, native to tropical America, have a robust body form, varying from the size of a guinea pig to that of a small beaver. The limbs may be modified for an arboreal, terrestrial, or semiaquatic life. Hutias (*Capromys*) are native to the West Indies while the larger nutria (*Myocastor*) occurs from Brazil to Chile. Nutrias have been widely introduced into Europe and North America.

Nutria. *Myocastor coypus* (Molina)

DESCRIPTION. The nutria is a large, stout-bodied rodent, with a long, scantily haired tail. This species is most apt to be confused with the muskrat, of native mammals. The tail is somewhat less than half the total length, and is *terete,* whereas that of the muskrat is laterally compressed. The skull is immediately separable from that of the muskrat by the infraorbital foramen, which is approximately the size of the orbit. This is the case also in the porcupine, but the infraorbital opening of the muskrat is tiny. Large individuals may exceed 20 pounds, although the average weight is about 12 pounds. The dense soft underfur is overlaid with yellowish brown guard hairs, coarsest on the back. The pale vibrissae are long and stiff. The hallux and first three toes of the hind foot are prominently webbed, while the fifth toe is free, and is possibly used in grooming. The teats are laterally placed in such a manner that the young can nurse while the mother is feeding in shallow water. Eight Louisiana individuals measured by George Lowery (1974) averaged: total length, 940 (837–1010) mm; tail, 344 (300–340) mm; hind foot, 131 (100–150) mm (Figure 6.73).

DISTRIBUTION.   Through accidental or purposeful releases, nutria have established themselves in wetland areas in many parts of the United States and Canada. Presently Michigan, Indiana, New York, Maryland, Virginia, Georgia, Florida, Alabama, Arkansas, and Louisiana have populations of these large rodents. The 1930s were considered the boom years for establishing nutria ranches. With the advent of World War II, nutria farming virtually collapsed. Ranchers released their stock. Promoters selling nutria as "weed cutters" were responsible for transplanting them throughout the southeast in the late 1940s. Although some populations have disappeared, nutria have adapted to a wide variety of conditions, and they persist in areas presumably unsuitable for their existence (Figure 6.74).

Figure 6.73.   Nutria, *Myocastor coypus.* Drawing by W. C. Dilger.

Figure 6.74. Location of established populations of the nutria, *Myocastor coypus.*

HABITS.   Nutrias are essentially animals of the freshwater marshes. They make their own burrows, or utilize those of other mammals. They frequently occupy muskrat houses, or make surface nests in dense vegetation.

The food consists of a wide assortment of vegetation, usually of an aquatic nature, but these creatures are adaptable and eat bark, Bermuda grass, clover, and assorted roots. Like the muskrat, nutrias often amass vegetation into feeding or resting platforms.

Breeding occupies the entire year. Gestation lasts about 130 days, when the four or five young are born in a well-furred condition with open eyes. They are able to swim shortly after they dry off. Miscarriage is unusually high in this rodent, about a third of the litters being lost during pregnancy.

Nutrias are capable of doing immense damage with their sharp incisors. In the coastal sugarcane land and rice fields, adjacent to Louisiana marshes, great loss is experienced. On the credit side, the soft fine underfur is valuable in the fur industry. In 1961, the average annual take of pelts was valued at over a million dollars. In more recent years, the sale of the pelts, used mostly in trim and lining, and of the meat for pet food, has brought trappers several million dollars annually. The first nutrias appeared in the Louisiana fur market during the 1943–44 trapping season; two years later 8,784 were taken, whereas in the 1972–73 season, trappers took over 1.6 million, yielding an income of over $7.5 million, making it the number one fur animal in Louisiana.

The alligator is the chief predator on the nutria in the southeast.

7

Carnivora

(Flesh Eaters)

The carnivores are a well-marked mammalian group, characterized by a dentition adapted for tearing and shearing flesh. The canines are large and pointed, projecting well beyond the other teeth. Most carnivores possess a specialized shearing tooth in the upper and lower jaw; this is known as the carnassial or sectorial tooth. The incisors are small. Usually five toes are present on each foot; these are invariably provided with claws, which are blunt in the dogs, but sharp and retractile in the cats.

Carnivores are native throughout the world with the exception of Australia. As their name implies, they are primarily flesh eaters, although many species, such as the bear, skunk, fox, and others eat large quantities of vegetation. They are less numerous than the rodents, on which they are largely dependent for sustenance.

Many eastern carnivores are important to the fur trade, and trappers catch large numbers of foxes, weasels, minks, skunks, and raccoons each year.

Key to the Genera of Eastern Carnivores

  A. Large molars with crowns of crushing type, not adapted for shearing.
      B. Skull large, more than 200 mm; molariform teeth, six in upper jaw, seven in lower jaw; color black; tail short; size massive ....*Ursus*
    BB. Skull medium, less than 150 mm; molariform teeth, six in upper jaw, six in lower jaw; tail bushy and ringed ....*Procyon*
AA. Large molars, generally with crowns not of crushing or grinding type.
    C. Large molars with shearing crowns; catlike.
      D. Tail short, less than length of hind foot; cheek and throat hair ruff present; molariform teeth, three in upper jaw, three in lower jaw ....*Lynx*
    DD. Tail long, more than length of hind foot; molariform teeth, four in upper jaw, three in lower jaw. ....*Felis*

CC. Rear molar teeth with crushing surface, tooth rows and facial part of skull relatively long, molariform teeth, six in upper jaw, seven in lower jaw; doglike.

    E. Size large, adult weight more than 20 pounds; upper incisors prominently lobed ....*Canis*

    EE. Size medium, weight less than 20 pounds; upper incisors not lobed.

        F. General tone reddish, tail tip white; cranial ridges forming a prominent V-shaped sagittal crest posteriorly. ....*Vulpes*

        FF. General tone grayish, tail tip black; cranial ridges lyrate, forming prominent U-shaped ridges on outer dorsal surface of skull ....*Urocyon*

CCC. Tooth row and facial position of skull relatively short; auditory bullae usually flattened; external form variable; size small to medium.

    G. Modified for aquatic life, toes fully webbed; molariform teeth, five in upper jaw, five in lower ....*Lutra*

    GG. Toes scarcely or not at all webbed.

        H. Molariform teeth, four in upper jaw, five in lower jaw; auditory bulla little inflated; color black and white.

          I. Upper outline of skull nearly straight; white stripes variable, the posterior ones at right angles to anterior stripes ....*Spilogale*

          II. Upper outline of skull convex; two white dorsal stripes variable in length and width................*Mephitis*

        HH. Auditory bulla inflated; color not black and white.

          J. Molariform teeth, four in upper jaw, five in lower jaw. Toes with long (15–25 mm) claws; size large; pelage long and coarse ....*Taxidea*

          JJ. Toes short; pelage soft and lustrous.

            K. Molariform teeth, five in upper jaw, six in lower jaw................................*Martes*

            KK. Molariform teeth, four in upper jaw, five in lower jaw ....*Mustela*

Coyote. *Canis latrans* Say

DESCRIPTION.   In appearance the coyote is not unlike a small police dog, but it has longer fur and a shorter, bushier tail. The color is variable; grayish tawny above, the dorsal hairs broadly tipped with black; head grizzled gray, ears pale brown; white on throat and belly; outer side of hind legs and feet fulvous. In the field, the coyote may be mistaken for the timber wolf. The latter animal, which is much rarer, can be recognized immediately by its much larger size, grayish white fur, and short tail held upright; the coyote has a yellowish cast and carries the proportionately longer tail held low. Adult male coyotes seldom measure more than four feet (1,200 mm); tail, 12–15 inches

(300–375 mm); hind foot, 7–8 inches (175–200 mm); weight, 20–50 pounds (9.1–22.7 kg). Exceptionally large individuals may weigh 55 pounds (25.4 kg). The females are smaller (Figure 7.1).

Presently one hears much about coydogs in the northeast. It is well established that dogs and coyotes mate and produce hybrids that exhibit characteristics of both species. Crosses between the coyote and several of the larger breeds of dogs, notably German Shepherds, Collies, and Airedales, have been taken in New York, and presumably other breeds are involved in these mixed populations. Only a small percentage of more than 200 New York animals examined in our laboratory, however, indicated a hybrid origin.

DISTRIBUTION.   The coyote is generally common in northern Wisconsin and northern Michigan, western Illinois along the Mississippi River, and northern Indiana, and occurs sporadically in nearly every eastern state, from Maine to Florida. Many of the coyotes which have been captured or reported in the eastern states have come as prospective pets in the autos of tourists, only to escape later from their owners, or perhaps to be set free when they attain large size. Others, particularly in the southern states, have been liber-

Figure 7.1.   Coyote, *Canis latrans.*

ated by fox hunters who have had coyote pups shipped to them instead of young foxes, to which as pups they bear a striking resemblance. According to Lawrence E. Hicks (personal letter, December 1940), stragglers have been seen in most of the counties of the western half of Ohio, and specimens have been taken in about a dozen counties. In the semirough country near the highest point in Ohio, near Bellefontaine, Logan County, coyotes have been well established since about 1935. About thirty-five or forty have been shot from this center alone, and Hicks has handled litters of pups.

The coyote is now generally distributed throughout New York and New England and is spreading southward, and in the southern states it is expanding eastward. We have shaded on the map (Figure 7.2) only areas where the species seems to be presently established.

HABITS. In spite of the encroachments of civilization, the adaptive coyote manages to maintain its numbers, and even appears to be increasing in all parts of its range. This is rather remarkable when we consider how this little prairie wolf has been trapped, hunted, and poisoned for many years, with a price on its head wherever it occurs. The coyote, like the red fox, is an inhabitant of the brush country, but is equally at home in the forests or farm land.

A symbol of the western prairies is the sharp yapping bark of the coyote at evening, or on the still morning air at sunrise. In the east, the bark is seldom heard except in heavily forested areas, perhaps because the coyote must make itself less conspicuous in a land where enemies abound.

The home is either a den made by the coyote or a remodelled fox, skunk, or badger hole. The coyote usually spends the day well concealed in a dense thicket or woods, for normally it is most active at night. It is not uncommon to see an individual abroad during the day.

Mating occurs in late winter. After a sixty-three-day gestation, the five to ten little blind pups are born. Both parents assist in feeding the young and make circuitous trips to the den to avoid attracting the attention of potential enemies. The family breaks up in the late summer or early fall, and the young hunt alone until late winter, when they are ready to pair and assume family duties themselves.

The coyote is accused of killing lambs, pigs, and poultry. Undoubtedly individuals are responsible for the destruction of some livestock, but their principal food consists of wildlife, chief among which are ground squirrels, pocket gophers, woodchucks, wildfowl, snakes, insects, and many plant foods. Fruits and berries are eaten during the summer months. Carrion of larger animals such as deer is important in the winter.

Hamilton (1974) examined 1,500 coyote scats from the Adirondacks, taken throughout the year. Mammals occurred in 78 percent, fruit 21 percent, insects 10 percent, birds 3 percent, with amphibians, reptiles, and green

Figure 7.2. Distribution of the coyote, *Canis latrans.*

grasses making up the remainder. The most important food was the snowshoe hare, occurring in 40 percent of the scats.

Man is the only serious enemy of the coyote. With dogs and gun, this little wolf is hunted relentlessly. Few dogs are swift enough to catch him, and the coyote is the equal of any dog of comparable size. The pelts are used in trimming ladies' garments, and, to a lesser extent, for scarves. Many are dyed various shades of gray, brown, and black. Western skins from the Rocky

Mountain areas earlier brought as much as $7.00 or more, but eastern hides of the best quality traditionally were worth only $2.00 or $3.00. However, in the past few years of rising fur prices, particularly for "long" fur, coyotes in Indiana have increased greatly in value. In the 1972–73 through the 1976–77 seasons, eastern coyote pelts brought average prices of $9.31, $10.82, $9.36, $8.46, and $15.08.

Gray Wolf. *Canis lupus* Linnaeus

DESCRIPTION. A large, broad-headed wild dog, *Canis lupus* is best characterized by the thick rich fur, powerful forelegs, and heavy muzzle. Much variation in color occurs, but the usual pattern is grayish, brownish gray, or brownish white, with middle of back heavily marked with black; underparts and legs pale rufous brown to yellowish white. Pelage is heavier and shaggier in winter. The young are dark gray to blackish when born. Other characters are the broad nose pad of the adult, measuring 30 mm or more across. Skull length is greater than 200 mm. Adult males have a total length of about 5.5 feet (1,625 mm); tail vertebrae about a foot (300 mm); hind foot, 10 inches (250 mm). Females are noticeably smaller. Large adult male wolves may weigh from 75 to 100 pounds (33–44 kg), while individuals occasionally attain a weight of 150 pounds (68 kg), although this must be regarded as exceptional (Figure 7.3).

DISTRIBUTION. The gray or timber wolf has been persecuted for so long that its range has decreased yearly. Where formerly this species ranged widely throughout the eastern states, persistent hunting, trapping, and poisoning resulted in its extermination in Pennsylvania, New York, and New England well before the close of the nineteenth century. There is no authentic record of its occurrence in these parts during the present century (Figure 7.4).

Its last stronghold in the United States east of the Mississippi is the more extensive and remote forest areas of Michigan's Upper Peninsula and northern Wisconsin. A. M. Stebler of the Michigan Department of Conservation informed Hamilton in December 1940 that the timber wolf population in the Upper Peninsula was probably somewhere between 100 and 150, and apparently some wolves still survive there and in adjacent Wisconsin. It is fairly definite that the last wolves have been trapped out of Douglas, Bayfield, and Sawyer counties, although there may possibly be a few in Iron and Vilas counties. There is a wolf population on Isle Royale in Lake Superior.

HABITS. Cunning and sly, the wolf is yet no match for man, and it has been all but extirpated in the east. Frequent newspaper accounts report the capture of a wolf in various eastern states, but these prove, on close inspection, to be coyotes or police dogs. East of the Mississippi this great wild dog is

Figure 7.3. Gray wolf, *Canis lupus.*

now restricted to the forested regions of northern Michigan. This was not true a century ago, when wolves roamed all through the forested east, taking fearful toll of the settlers' livestock. Great hunts were then organized to reduce these pests, but bounties, poison, and traps were not particularly effective. The eventual extermination of the wolf over its primitive range was in large measure due to the settlement of the country and the attendant harassment of the species.

Wolves mate in the late winter, and after a gestation of sixty-three days a litter of five to eleven pups is born. Their growth is much like that of puppies, and the young commence to hunt with the parents in late summer. The family remains together during the first winter.

Two families may join to form a pack of twelve to fifteen young. In Ontario, at least, the pack has a large area over which it hunts, returning at quite regular intervals. The circumference of the hunting range may cover several hundred miles.

Figure 7.4. Distribution of the gray wolf, *Canis lupus.*

During the winter season the gray wolf subsists principally upon deer. If the entire deer is not consumed by the pack, they return at a later date to feed on the carcass. Northern trappers utilize this trait by poisoning the carcass or placing traps at advantageous points about the dead animal. In the spring and summer the snowshoe hare, ground squirrels, mice and other small mammals, birds, frogs, snakes, crayfish, molluscs, and even insects are eaten.

Wolf pelts were previously used as scarves and trimmings, but this species is now granted protection in the east.

Red Fox. *Vulpes vulpes* Linnaeus

DESCRIPTION.   The red fox scarcely needs description. It is the size of a small dog, with pointed nose, large prominent ears, and long bushy tail. Upperparts are reddish yellow, mixed with black-tipped hairs in the median line, the rump grizzled, reddish hairs mixed with white- and black-tipped hairs; feet black; the bushy tail mixed with black, a black spot at the base and the tail tip white; underparts, cheek, and inner side of ear whitish; back of ear black. Color varies with the seasons, the fur full and lustrous in winter pelage; faded, pale, and relatively short in summer. Various color phases occur, the best known of these being the silver fox, with melanistic coat frosted with white, particularly about the head, shoulders and rump, and the cross fox, whose pelt is mixed with gray and yellow. These color phases may occur in the same litter with normal reds. The skull has prominent temporal ridges which unite to form a sagittal crest; upper incisors lobed. Measurements of eight New York, Michigan, and Wisconsin adults average: total length, 972 mm; tail, 371 mm; hind foot, 163 mm; weight 8–12 pounds (3.6–5.5 kg), the males larger than the females. Extreme weights may approach 13 or 14 pounds (5.9–6.3 kg), although the average adult weighs but 10 or 11 pounds (4.5–5 kg) (Figure 7.5).

DISTRIBUTION.   The red fox occurs over most of eastern North America, from Maine to Wisconsin, south to northern Georgia and Mississippi, though it is less common southward (Figure 7.6). The red fox of the United States was long recognized as a separate species, *Vulpes fulva,* but recent studies have indicated it to be of the same species as the old world red fox, *Vulpes vulpes.* Red foxes have been introduced into the southern states for purposes of sport; some of these may prove to be *Vulpes regalis.*

HABITS.   Few animals occupy such diverse habitats as our red fox. It prefers the rolling farm land, mixed with sparsely wooded areas, marshes, and streams. It also occupies the borders and open areas in heavily forested regions or within the very limits of the great metropolitan areas of Boston and New York.

Figure 7.5. Red fox, *Vulpes vulpes* (top), and gray fox, *Urocyon cinereo-argenteus.*

Figure 7.6. Distribution of the red fox, *Vulpes vulpes.*

Best known for its sagacity and cunning, the red fox continues to maintain itself and even increase in spite of constant persecution. Although it may live for years in quite thickly settled areas, the sight of a hunting fox should be recorded as a notable event in the naturalist's notebook. While Reynard is usually more active by night, he is abroad also by day, and occasionally may be observed assiduously hunting mice in the meadows. If the observer be quiet and not too distant, squeaking will draw the attention of the fox, and, if the wind is favorable, the beast may be drawn to within a rod or two of the observer. Whitaker once saw a fox hunting in a meadow in New York State. He fell to the ground and lay still. The fox took a big circle and came up behind him to within sixty feet before making off.

As winter wanes, the foxes begin their nocturnal barking; this is the mating season. Fifty-one days after mating, four to ten young are born, usually in late March or early April. The pups do not come to the den entrance until they are about five weeks old. There is hardly a more attractive sight than the youngsters about the den entrance, rolling and tumbling in play, while the mother, ever alert, watches over them. The father is a devoted parent, and shares the burden of feeding the growing pups. The den site is chosen with some care, and the excavated dirt is scattered so that no telltale mound is readily visible. Often a woodchuck burrow is remodelled. The den is often in a wooded slope and not infrequently in an open field. Family ties are broken in August and the young foxes disband to seek a hunting ground, and in the following winter they mate.

The fox is no epicure; it feeds on grass and geese alike. Its chief food during the long winter appears to consist primarily of mice, rabbits, such birds as it can secure, carrion, frozen apples, dried berries such as grapes, and offal thrown up by the tides. In the spring and summer months woodchucks, poultry, rabbits, small rodents, birds, snakes, turtles and their eggs, an occasional young fawn, raspberries, and blackberries provide a varied diet, while in the fall berries, particularly wild cherries and grapes, grasshoppers, and the ubiquitous mice are mainstays.

Fox hunting is considered great sport, whether by the use of hounds, on horseback with a pack, or still-hunting when snow covers the fields. Many are taken in traps, and although it is considered difficult to take one in this manner, experienced trappers may take a hundred foxes in a season. Their success can be largely attributed to the use of a good scent, which will attract a fox to the lure from a distance of half a mile. The thick handsome pelt was formerly worth from $10 to $20 or more, but in the 1940s and 1950s prime skins were worth not more than a few dollars. In recent years, however, long furs have again become popular and fox pelts are now valuable. Average prices of red fox pelts in Indiana were $4.95, $4.13, $7.42, $16.49, $18.84, $18.69, $19.09, and $34.26 in the trapping seasons 1969–70 through 1976–77, respectively. As one might expect, this price increase brought about a great increase in the amount of fox hunting. The numbers of foxes taken in those years were 4,116; 2,939; 2,920; 5,843; 12,421; 10,269; and 14,516. In the fall of 1976, New York pelts brought $50, with 100,000 pelts being sold. The tremendous popularity of the silver fox (a melanistic phase of the red fox) once resulted in an extensive growth of the fur-farming industry, but presently the number of foxes raised for their fur is negligible. Fox pelts are used almost solely for trimming and scarves.

Gray Fox. *Urocyon cinereoargenteus* (Schreber)

DESCRIPTION.   In appearance, the gray fox is slightly smaller than the red fox, with a shorter muzzle and shorter legs. Color is grizzled gray above, the hairs banded with black and grayish white; inner sides of legs, sides of belly, neck and the band across the chest reddish brown; the rest of the underparts are white; the sides of nose and underjaw is blackish; the tail has a concealed mane of stiff black hairs on its upper side near base, which can be felt by rubbing one's hand along the tail toward the body. The skull has prominent lyrate temporal ridges, which do not join to form a prominent sagittal crest as in the red fox, and upper incisors are not lobed. The length of the skull is 124–130 mm. Average measurements of ten adults from Maryland, North Carolina, and Virginia are: total length, 970 mm; tail, 347 mm; hind foot, 143 mm; weight, 7–11 pounds (3.2–5 kg). Hamilton has seen a 14-pound (6.3 kg) New York individual, and J. E. Hill has recorded a 19-pound (8.6 kg) Massachusetts specimen. Southern specimens seldom exceed 10 pounds (4.5 kg) (Figure 7.5).

Gray foxes in the northern parts of the range are larger than those in the south. The skull is larger: basal length of skull, 126 mm; zygomatic breadth, 73 mm; palatal length, 63 mm; front of canine to back of last upper molar, 56 mm. Individuals from Wisconsin are larger; much of the gray is replaced by yellowish gray; the reddish areas of neck sides and legs are more rich and more ferruginous; sides and underpart of tail are more ferruginous, the tail tip duller. The skull is large; audital bullae are very much smaller and flatter than in the typical form. Gray foxes from Florida are smaller than *cinereoargenteus* of the middle states, with relatively shorter hind foot, tail, and ears, and harsher pelage. Skull characters are not appreciably different. The fulvous coloring of the breast and belly is quite pale and the almost total absence of white in these parts is characteristic. Two adults from central Florida average: total length, 905 mm; tail, 285 mm; hind foot, 125 mm.

DISTRIBUTION.   The range of the gray fox includes all of eastern United States, from New Hampshire and Wisconsin south to the Gulf (Figure 7.7).

HABITS.   The gray fox is typically a southern or western species, and has invaded the extreme northern limits of its range only within recent years. It is found in wooded areas, swamps, and the hammocks and pine woods of the southern states, but does not take so kindly to the farm lands as does the red fox. This species is an adept climber, often taking to trees when pursued by hounds. In climbing, the fox may leap cautiously from limb to limb, using its stout nails in the manner of a cat, or it may actually climb in the manner of a bear.

A hollow tree or log may provide a den site; northern gray foxes rely on a burrow in the ground, but these are seldom found in the open fields as those of

Figure 7.7. Distribution of the gray fox, *Urocyon cinereoargenteus.*

the red fox occasionally are. Less cunning than its red cousin, the gray fox usually takes to cover after a short run by the hounds, repairing to this underground retreat or climbing into the crotch of a suitable tree. It often frequents the borders of swamps or lowlands, and may venture into the marsh for ducks and muskrats.

The two to seven (but usually three to five) young foxes are born in March or April, depending upon the latitude. Gestation is about fifty-three to sixty-three days. The young are served by the solicitous parents into the early summer, and they continue with the mother until they are fully capable of maintaining themselves.

Perhaps the most important food of the gray fox is the cottontail rabbit. It is not surprising that the major flea of this species is *Cediopsylla simplex,* a flea of cottontail rabbits. From the few stomachs of this fox that we have examined, we are inclined to believe that it eats more birds than the red fox does. Small mammals, particularly field mice, deer mice, cotton rats, wood rats, and shrews provide many meals, while snakes, turtles and their eggs, lizards, insects and their kin, apples, beechnuts, corn, peanuts, grapes, hickory nuts, persimmons, carrion, wild cherries, and grasses are additional foods.

Other than man, the gray fox has few enemies. Hamilton found the remains of one in the stomach of a large Vermont bobcat. The coarse gray fur is widely used for collars and trimmings, but its quality does not match that of the red fox and the trapper receives less than for red fox hides. Gray fox pelts in Indiana brought an average of $10.48 and $19.51 in the 1974–75 and 1975–76 seasons, whereas red fox pelts averaged $19.09 and $34.26 those years. New York pelts brought the trapper $.75 in the 1930s, but fur buyers paid $25.00 for pelts in 1976. The gray fox undoubtedly kills a considerable number of game birds, but the destruction of numerous rodents and rabbits, particularly in areas where these are harmful to crops, should balance the budget so far as man is concerned.

Black Bear. *Ursus americanus* (Pallas)

DESCRIPTION.   The black bear is the only bear occurring in the eastern United States and needs little description. It is a large, heavy, thick-set animal with rounded ears, coarse dark fur, and a very short tail, so small as to be nearly concealed in the fur. In summer, individuals vary in color from cinnamon brown to blackish, but in winter the coat is long, thick, and glossy black. The face and nose are tinged with tan; occasionally a white spot appears on the breast. Adults from the northern parts of the range measure about 1600 mm (64 inches) and stand slightly more than two feet at the shoulder. Full-grown specimens weigh 200–300 pounds (91–136 kg) but exceptional individuals may weigh 400 pounds (182 kg) or more. The skull has a basilar length of 250 mm, zygomatic breadth 170–180 mm, and facial angle 35 degrees; molar tooth row 75–80 mm (Figure 7.8).

DISTRIBUTION.   The black bear formerly ranged over all the forested portions of the eastern United States, but with encroaching civilization it has become increasingly scarce. It is, however, presently abundant in parts of Maine, northern New Hampshire and Vermont, western Massachusetts, much

Figure 7.8. Black bear, *Ursus americanus.*

of New York and Pennsylvania, northern Michigan and Wisconsin, the mountainous parts of West Virginia and Virginia, the Smoky Mountains of North Carolina and Tennessee, and possibly the Dismal Swamp (Figure 7.9). In the deep south, the species still holds forth in southern Mississippi and Louisiana, most of Florida, and southern Georgia.

HABITS. The black bear is today largely restricted to the wilder portions of the east—the secluded northern forests or the almost inpenetrable southern swamps where man and his dogs do not habitually trespass. Individuals frequently visit populated agricultural regions, but do not tarry long in such places. Bears are great travelers, sometimes covering many miles in a single night. The minimum estimated (via radio telemetry) home range size of five bears in Louisiana averaged about twenty-one square miles and ranged from about six to sixty, with that of the males averaging much more than that of the females. Three of the bears became inactive for periods of 74 to 124 days, and one gave birth during this torpor. The bears were found to den in a variety of situations, such as tree cavities (often with an opening far above the ground), hollow logs, and road culverts (Taylor 1971).

Bears are omnivorous, selecting a wide choice of foods as we might expect any carnivore to do. These big animals are inordinately fond of fruits, consuming great quantities of blueberries, blackberries, and shadberries. Mice, insects and their grubs, stranded fish, and other small fry are eagerly taken. Occasionally they destroy sheep and pigs, but their usual fare is that which they secure far from the farmlands. In the fall, bears eat quantities of

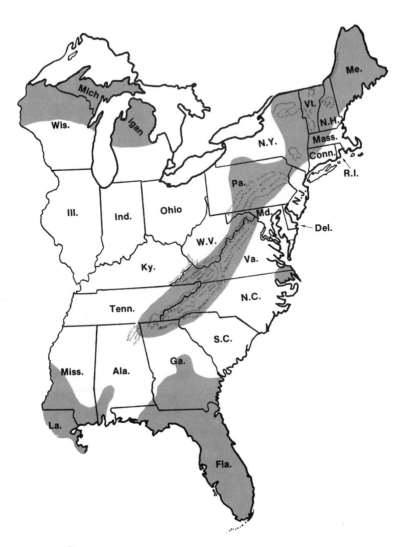

Figure 7.9. Distribution of the black bear, *Ursus americanus.*

acorns and other nuts and occasionally the leaves of hardwoods. In Florida cabbage palmetto berries and buds are important foods. A large bear will climb to the top of a cabbage palmetto, put his arms around the top and sway back and forth until the big bud loosens, then fall to the ground on his back, the tender cabbage secure in his paws. Other foods include the large boll berries, ants, grubs, and honey, for which the black bear has a hearty appetite.

With the approach of winter, they become extremely fat and are thus well equipped for the long dormant winter season. When snow flies the females seek some shallow cave, large hollow log, windfall, or even the open floor of a sphagnum bog, sheltered only by the overhanging boughs of a spruce or hemlock. The males retire later to their winter quarters. Even in Florida bears are said to ''hole up'' during the coldest weather. No bear truly hibernates, for its temperature does not fall nor is respiration greatly retarded as one finds with true hibernators.

The female gives birth to her one to three (normally two) cubs in late January or February, after a seven-month gestation. The cubs are remarkable for their diminutive size; the new-born offspring of a 200-pound adult scarcely exceeds 8 ounces and is no larger than a guinea pig. A period of delayed implantation of the blastocyst explains the small size of the cubs in relation to the long gestation. The youngsters remain with the parent through the summer and fall and may continue with her the following winter. Bears ordinarily have young every other year.

Other than man, black bears have few enemies. Perhaps their greatest fear is the ever-present terror of fire, from which even the swiftest of forest mammals can seldom find escape.

The fur has little value, but the secretive habits and large size make the black bear a favorite with hunters. Many hundreds are killed each year in the eastern United States. The flesh is highly esteemed.

Eastern Raccoon. *Procyon lotor* (Linnaeus)

DESCRIPTION. The raccoon is a medium-sized animal. Its pelage is long and thick, with a bushy ringed tail; muzzle slender, black facial mask, ears prominent and rather pointed; the front and hind feet with five toes, armed with prominent climbing claws; soles of the hind feet naked. Color is variable, the general tone is dull yellowish gray or grayish brown, darker on the back, many of the hairs black tipped. The face is whitish, a prominent black facial mask running from cheeks to eyes, a black streak on forehead, buffy to grayish black; back variable, grayish black to yellowish black; general tone of underparts dull gray to yellowish gray, the guard hairs whitish or yellowish white; tail yellowish or gray, with four to six black or brown rings, not well defined below; feet white. The skull is broad and rounded; molars broad and tuberculate; palate extending back of last molar for about 18 mm, bullae slightly flattened and extending laterally in a tubular auditory meatus (Figure 7.10).

Northeastern raccoons measure 700–830 mm in total length; tail 220–250 mm; and hind foot, 105–120 mm. Large individuals will weigh up to 22 or 24 (10–10.8 kg) pounds, but adults commonly weigh about 15 to 18 pounds (6.8–8.2 kg). Individuals from Wisconsin and upper Michigan are the largest and darkest of the eastern raccoons, with long full soft pelage suffused with ochraceous buff; skull large and massive, braincase more tapering anteriorly;

Figure 7.10. Eastern raccoon, *Procyon lotor.* Photo by W. J. Schoonmaker.

condylobasal length usually 120 mm or more. Length of adults is 700–880 mm. Weight is rarely up to 30 pounds (13.7 kg). We cannot give credence to reported weights of 56 and 62 pounds for Wisconsin raccoons.

Raccoons become progressively smaller to the south. Individuals from Florida are similar to typical *lotor* but somewhat smaller, with considerably longer tail; front and hind feet larger; legs longer, the animal when on all fours standing much higher; ear not so pointed; color more yellowish above with distinct and bright shoulder patch, in many specimens deep orange-rufous; pelage short and harsh. The Florida coon, in the flesh, is a more "leggy" creature than *lotor* of the north. Measurements of ten adults from Florida and southeastern Georgia average: total length, 812 (751–892) mm; tail, 262 (242–286) mm; hind foot, 122 (110–129) mm; weight, 10–15 pounds. Raccoons of extreme southwestern Florida and the keys seldom exceed 8 pounds (3.6 kg).

The chain of keys bordering the southwest coast of Florida, known as the Ten Thousand Islands, and the great series of islands from Biscayne Bay to

Key West, commonly known as the Florida Keys, have by virtue of their isolation and exposure given rise to several well marked and depauperate insular forms. In the nearly impenetrable mangrove swamps, where fresh water streams are unknown, these small raccoons are extraordinarily numerous. E. W. Nelson states that single trappers have taken 800 coons in a season when prices were high.

DISTRIBUTION. The genus *Procyon* is widely distributed in the eastern United States, from Maine to Wisconsin, and south to the mangrove swamps of the Florida Keys and the vast bayou country of Louisiana (Figure 7.11).

HABITS. The raccoon is found wherever suitable conditions of woods, swamps, and streams provide acceptable food and den sites. In the north it inhabits forested regions, but in the south it favors the swamps and mangrove thickets.

The raccoon is an accomplished climber, its sharp claws enabling it to ascend a tree of any size with remarkable celerity. It descends either head or tail first. It usually makes its home in a hollow tree, although the large fissures in cliffs may provide a suitable den site. Even woodchuck holes are utilized for sanctuary, although we have never known the raccoon to use one of these as a breeding chamber.

The raccoon is a creature of the night, commencing to forage along the creeks and streams after sunset. The muddy banks and bottoms of small streams through wooded areas in the midwest are often crowded with raccoon tracks, in low water of late summer and fall. The raccoons make their way along the stream, visiting each pool of water in search of crayfish, frogs, fish, or other easy prey. Several times we have been afield at night and heard a raccoon splashing through the water as it hunted by wading along the shore. At dusk one can often see whole families of raccoons moving about in den trees. Occasionally an entire family will be surprised away from its den site during the late afternoon or early morning hours, but such encounters are infrequent. In suburban areas, large drainage tiles, chimneys, and other man-made structures are used for sanctuary.

In the northern part of its range, the raccoon takes the approach of winter as a signal to retire to some warm retreat, to wile away the most severe months in sleep. Such dormancy is not akin to hibernation, for the metabolism is not lowered as with the woodchuck and other true hibernators. During mild spells of December and January it is up and about, searching for other life which is active during the milder spells of midwinter.

In late January or early February, the mating urge arouses the raccoon, and after a gestation period of sixty-three days, the three to six young are born, usually in early April. At birth they are well covered with fur, and they soon acquire the markings of the adult, but their eyes do not open until they are nearly three weeks old. The mother is a devoted parent: she teaches the

Figure 7.11. Distribution of the eastern raccoon, *Procyon lotor.*

young to climb, to hunt, and to practice the crafty habits so necessary to maintain them against man and his dogs. The family remains together into the late fall, often well into winter. The youngsters weigh from 7 to 12 pounds (3.2–5.4 kg) by early winter, and mate before they are a year old.

The raccoon is an omnivorous creature, feeding on a variety of plant and animal matter. It is particularly fond of crayfish, various fruits and berries, nuts, grains, and insects and other invertebrates. In the summer, the raccoon will strip the ears of growing corn, and later, like many other mammals, it feasts on the myriads of crickets and grasshoppers which swarm at this season. It is a widespread belief that the raccoon washes all of its food before eating, whence its name, *lotor,* a washer. Although captives usually dip their food in water, this is by no means an invariable trait of their wild kin, for much of the food is secured in regions quite inaccessible to water.

The season of peril is the fall, when hunters and their dogs take to the woods on November nights throughout much of the range of the raccoon. In addition to the exciting sport of pursuing the coon through swamps and beech ridges, the hunters gain a further reward through the sale of the pelt. The fur is thick and durable, though coarse, and is widely used in the manufacture of short coats and trimmings. The average price of 450,000 raccoons in New York in 1976 was about $20.

Key to Weasels of the Genus Martes

A. Orange on throat and chest. Skull rounded behind. Greatest length of skull less
   than 95 mm . . . . *Martes americana*
AA. No orange. Skull angular behind. Greatest length of skull more than 95 mm . .
. . . . *M. pennanti*

Marten. *Martes americana* (Turton)

DESCRIPTION.   An arboreal weasel, the marten is about two-thirds the size of a house cat, with soft lustrous pelage, broad rounded ears, long lithe body, short legs provided with sharp claws, and a noticeably bushy tail. Coloration generally yellowish brown, richer and darker above. The throat and chest are conspicuously lighter than above, being orange or ochraceous buff. There is not a marked seasonal difference in color, although the summer pelage is thin and harsh. Males have a prominent elongated belly gland. Males average: total length, 525–600 mm; tail, 185–205 mm; hind foot, 80–85 mm. Females are smaller than the males. Martens weigh 750–1130 grams (1.6–2.4 pounds) (Figure 7.12).

DISTRIBUTION.   The marten was originally distributed throughout the spruce forests of eastern United States from Maine to Virginia and west to Wisconsin. Relentless trapping has extirpated it over most of this area, so that it is positively known today only from Maine, northern New Hampshire, northern Vermont, and the Adirondack Mountains of New York, where it exists in some numbers. An acquaintance of the authors took fourteen martens near Wanakena, New York, during the 1932 season. Possibly it still occurs in the wilder Berkshire Mountains of western Massachusetts (Figure 7.13).

Figure 7.12.   Marten, *Martes americana.*

HABITS.   The marten is an inhabitant of the cool northern spruce and balsam forests. It seldom ventures from this zone, nor does it need to, for the red squirrels, upon which it largely feeds, abound in this habitat likewise.

The habits of the marten are little known, due in a measure to its relative scarcity and to the nature of the habitat which it occupies. It retreats before the advance of civilization, yet is of an inquisitive and curious nature, falling easy prey to traps. Much time is spent among the trees, where it pursues and captures the frenzied red squirrel, It seeks the ground to hunt mice and chipmunks and secure other edibles.

The nest is made in some rotten snag or hollow tree, and the marten's small size enables it to utilize the holes made by the big pileated woodpecker. In such a retreat a nest of leaves is prepared, and here the marten spends the greater part of the day in sleep, emerging at night to hunt. It is often active in the day, however, and trapper friends of mine have seen martens on the ground or pursuing squirrels aloft. Its color and general physiognomy suggest a red fox, particularly so when the marten is seen on the ground.

Mating occurs in late July or August, but shortly after the ova have been fertilized development is arrested for several months. Thus implantation of the embryo to the uterine wall is delayed and foetal development is not resumed until midwinter. The three to five young are usually born in mid-April after a gestation of from thirty-one to thirty-three weeks. At birth they are blind, helpless and covered with fine yellowish hairs. The eyes do not open until the

Figure 7.13. Distribution of the marten, *Martes americana.*

youngsters are nearly six weeks old. Adult weight is attained in about three months, when the females weigh about 1.6 pounds and the males about 2.5 pounds.

While it appears that the chief food of the marten consists of squirrels, mice, such birds as it can capture, and other small animal life, the marten is known to eat the berries of the mountain ash and other wild fruits.

Among enemies may be listed its cousin, the fisher, and all the larger predators which inhabit the same territory. Because of his relentless pursuit of furs, man of course is the mortal enemy of the marten. This species is easily trapped, even during periods of heavy snow. The fine quality of the pelt, and the high price it commands, have caused a great reduction in its numbers. Lumbering operations likewise have dealt it a severe blow, for much of the territory which was formerly occupied by the marten has now been destroyed. Currently there is no open trapping season for this species in the east.

Fisher. *Martes pennanti* (Erxleben)

DESCRIPTION.   The fisher is a large, dark brown weasel, about the size of a fox, but with shorter legs and less prominent ears. The face is somewhat pointed, the broad ears rounded; neck, legs, and feet stout, the latter with strong stout climbing claws; tail tapering and bushy. The fur is dense and soft, although much coarser than that of the marten. Color is dark brown to black-ish brown, the face and shoulders heavily streaked with gray and paler brown; underparts nearly black, a few white patches on neck and throat. It is paler in summer. Measurements of three adult males from New York and Maine aver-age: total length, 970 mm; tail, 360 mm; hind foot, 126 mm; weight, 8–12 pounds (3.63–5.5 kg). Females are considerably smaller than the males (Figure 7.14).

DISTRIBUTION.   At the present time breeding populations of the fisher are probably restricted in the east to northern Maine and the Adirondack Mountains of New York (Figure 7.15). The fisher has staged a dramatic

Figure 7.14. Fisher, *Martes pennanti.*

increase in New Hampshire, particularly the southern part of the state. Poole (1932) found tracks of a fisher in the Blue Ridge Mountains in Berks County, Pennsylvania, during the winter of 1931. He has recorded a specimen captured at Holtwood, in Lancaster County, Pennsylvania, in 1921. Populations are sufficiently large in New York and Maine to justify an open trapping season. Adirondack fishers were introduced into northern Wisconsin in 1956, and a number were released in the Catskill Mountains of New York in 1976.

HABITS.   In spite of continued hunting for its valuable pelt, the fisher still thrives in some numbers in the wilder forested mountain regions of Maine, extreme northern New Hampshire, and the Adirondack Mountains of New York. It is a dweller of the spruce forests, at home only where the wild country has not been invaded by man.

Figure 7.15. Distribution of the fisher, *Martes pennanti.*

In the Adirondacks it is still tolerably common and is occasionally seen in the daytime. It is an agile climber, pursuing even the red squirrel among the snow-covered conifers. The fisher is known to be a great traveler, frequenting a virtual circuit and denning in suitable places as it makes its rounds at regular intervals. It is known to swim sizable lakes, and is often found in timbered northern bogs and swamps.

The one to four young are born in late April in a snug nest situated in a cavity in a large tree, or in a log or rock ledge. Shortly after the young are born, the mother leaves the den to mate, for the gestation period is prolonged, lasting 348 to 355 days. The young grow slowly and their development is much like that of the young of the marten.

Its name is a misnomer, for the fisher does not prey extensively on fish. It is an inveterate enemy of the porcupine but is not always immune to the sharp quills, and the fisher has been found starving, its head and body riddled with quills. The fisher is suspected of attacking its lesser cousin the marten, in addition to foxes, skunks, and even the deer. Deer are particularly vulnerable to the attack of the fisher during periods of deep snow, when flight is impeded by drifts. In addition, rabbits, mice, and the lesser fry which abound in its northern home are frequent victims.

Man is its only important enemy, for the prized pelt is highly sought by trappers. Prime New York pelts sold for $100 in 1976. The fisher is easily caught and is not infrequently taken in traps set for the marten. Recently many states have given the fisher protection, and this respite may further its increase.

Key to Weasels of the Genus Mustela

Males are much larger than females in this genus, creating more overlap in size ranges than otherwise would be expected in animals of such differing sizes.

A. Tail not black tipped.
   B. Size small, less than 300 mm total length. Skull less than 40 mm long. Tail about one inch in length .... *Mustela nivalis*

  BB. Size larger. More than 300 mm total length. Skull more than 40 mm long. Tail much more than one inch . . . .*M. vison*

AA. Tail black tipped.

    C. Total length usually 300 mm or more, at least in northern parts of our area, where it occurs with *M. erminea* (except in small females). Tail generally more than 85 mm. Postglenoid length of skull (from point of articulation of lower jaw to posterior end of skull) less than 47 percent of distance from occipital condyles to tip of anterior incisor . . . .*M. frenata*

    CC. Total length generally less than 300 mm, except in occasional males. Tail generally less than 85 mm. Postglenoid length of skull more than 47 percent of this length . . . .*M. erminea*

Ermine. *Mustela erminea* Linnaeus

DESCRIPTION. The ermine or short-tailed weasel is a small, slender predator, scarcely larger than a chipmunk and built much more slimly. Its smaller size and shorter tail serve to distinguish it from the larger long-tailed weasel. The soft but rather thin fur is a uniform chocolate brown above, slightly darker on the crown of the head. The under parts are whitish, often stained with a pronounced yellow. Occasionally a small brown spot or two occurs on the throat or chest (Figure 7.16). The weasel dons a pure white coat in winter, retaining only the jet-tipped tail. This species probably becomes white in winter throughout its range, for of several thousand winter pelts we have examined, not one has been brown.

The measurements of thirty-one adult males average: total length, 272 (251–295) mm; tail, 70.8 (65–80) mm; hind foot, 35 (32–38) mm. Fifteen adult females average: 236 (194–255)mm; tail, 55 (44–54) mm; hind foot, 29 (28–31) mm. Males average 80 grams, females 54 grams. Minnesota specimens are heavier than eastern individuals.

DISTRIBUTION. This little weasel is a boreal species, being most numerous in the coniferous forests of northern United States and Canada. It is a wide-ranging form, occurring from Pennsylvania to Alaska. In eastern United States the short-tailed weasel occurs throughout most of New England, New York, and Pennsylvania, but is uncommon in coastal regions. Possibly in the Allegheny Mountains it extends south of Pennsylvania, but such records are lacking. The species occurs westward in most of Michigan and Wisconsin, but not to the south in Illinois, Indiana, or Ohio (Figure 7.17).

HABITS. This is an animal of the forest, and it prefers the deep spruce stands of the north. In the southern part of its range, it occupies a habitat similar to that of the larger long-tailed weasel, and is often found in brushy fields bordering cultivated areas.

Figure 7.16. Long-tailed weasel, *Mustela frenata,* male (above) and female (upper center), and the ermine, *Mustela erminea,* male (lower center) and female (bottom).

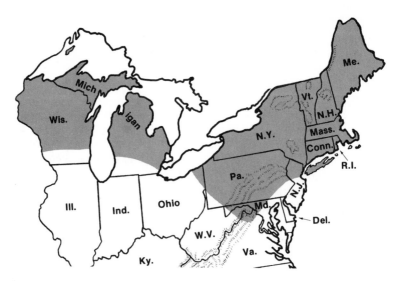

Figure 7.17. Distribution of the ermine, *Mustela erminea.*

The ermine, although a fairly common animal, is seldom seen by the naturalist. It seems to prefer the hedgerow or stone wall where it can elude its many enemies and catch the small mammals and birds which are attracted to these coverts.

All summer the weasel is garbed in a dark brown coat, but at the approach of winter a remarkable change occurs. A few white hairs appear on the tail and head; these rapidly increase in number until the animal has a curiously salty appearance; finally, the brown coat disappears entirely, being replaced by a pristine coat of white. This moult occupies from three to five weeks, the time varying with the individual. Nor do the animals by any means don this "ermine robe" at the same time. Individuals taken on the same date in early November may show only incipient change, or may have almost completely lost their summer pelage. The presence or absence of snow is certainly not a primary cause of the change, for many individuals become wholly white before the first snowfall. Hamilton has seen specimens with much brown fur remaining in early December. By early March a few brown hairs make their appearance, and with the swelling buds of April, the weasel has again assumed its summer coat of brown.

The weasel seldom makes its own home; it prefers to occupy the deserted chambers of a chipmunk, the cavity beneath some sizable stump, or a shallow excavation beneath a pile of boulders. The nest is a large loose structure, usually composed of the fur and feathers of its victims.

The ermine feeds chiefly upon mice, chipmunks, shrews, and other small mammals. It will eat frogs, lizards, small snakes, and many kinds of insects, and has been known to feed quantities of earthworms to its young. It destroys such small birds as it can capture. One-third of the food by bulk of 191 New York specimens which Hamilton examined contained remains of field mice. These weasels do not always confine their hunting to such small prey, for they can and do capture small cottontails.

All the evidence from a study of the reproductive organs of these weasels suggests that they mate in the early summer, and the fertilized eggs, after undergoing a short development, remain quiescent for several months. Embryonic development continues in the late winter, and the four to nine young are usually born in mid-April. They are tiny creatures at birth, sparsely covered with fine white hairs. When the baby is about three weeks old, a prominent mane appears on the neck, and this persists for several weeks while the fur on the remainder of the body grows in. The eyes do not open until the young weasel is five weeks old. Young males at seven weeks are larger than their mother.

The male weasel assists in bringing food to the young during their infancy. There is much evidence that weasels remain paired throughout the year. They are surely not the unsociable creatures that they are usually thought to be.

Owls, hawks, predatory mammals, including the house cat, and large snakes are their principal enemies.

The pelt value is small, but the large number taken by the northern trapper justifies a line of weasel traps for these handsome little animals. Weasels which are changing from brown to white are known in the fur trade as "graybacks" and are worth but a few cents. The weasel's greatest value lies in its role as a rodent destroyer. It is a valuable asset in any agricultural community.

Least Weasel. *Mustela nivalis* Linnaeus

DESCRIPTION.   The least weasel is at once recognized by its diminutive size and *very short tail,* this organ scarcely exceeding an inch in length. These little weasels have a dark walnut-brown summer pelage, except for the white underparts and toes. The tail tip contains a few black hairs, but the usual pronounced black tip is lacking. White winter specimens may have a few black hairs in the tail tip. Total length, 150–200 mm; tail, 20–35 mm; hind foot, 19–21 mm. Females are slightly smaller than males. Two male adults from Pennsylvania which we weighed were 46 and. 37 grams respectively. Thirty-six Indiana individuals ranged from 21.5 to 67.9 grams. This species might be confused with a female *M. erminea,* but the inch-long tail and the small skull, totaling not more than 32 mm in length, are sufficient to characterize it (Figure 7.18).

E. L. Poole

Figure 7.18. Least weasel, *Mustela nivalis.*

DISTRIBUTION. The least weasel occurs from mountainous western Pennsylvania southward into the mountains of North Carolina. Westward it occurs throughout much of Ohio, northern Indiana and Illinois, and much of Wisconsin. Examination of the pelts handled by local fur buyers will probably prove it to be more widely distributed than hitherto recorded (Figure 7.19).

HABITS. This tiny weasel is found in the deep forests of the high Alleghenies as well as the mid-west agricultural lands.

The least weasel is the smallest American carnivore, weighing but one ten-thousandth as much as the largest of our carnivores, the great Alaskan brown bear. Its diminutive size is scarcely calculated to draw the attention of the eye. As a result, its presence in the community is seldom known until one is trapped. It does not appear to be common in any part of its range. In Pennsylvania, where thousands of weasels have been trapped for bounty, relatively few of these little weasels have been taken.

Few Alleghenian least weasels appear to change to a white winter coat. Several white ones have been recorded from Pennsylvania, Ohio, Indiana, and elsewhere, but the majority acquire a slightly paler brown coat with the approach of cold weather. The manner of change and time required have not been described.

Little is known of the home life of this weasel. Nests have been found beneath corn shocks, in shallow burrows bordering streams, and in similar places. The few nests that have been examined were composed of grasses and mouse fur.

The breeding habits differ from those of the larger weasels. Young with unopened eyes have been discovered in midwinter, while lactating females and nest young have been found in Pennsylvania during October, January, and February. These litters numbered from three to six young. The female parent was always in attendance.

Gary L. Heidt (1970) has studied reproduction and development in this species. *Mustela nivalis* has two or more litters per year with one to six young per litter. Delayed implantation is not exhibited. The young are born after a gestation of about thirty-five days, with the newborn weighing about 1.4 grams each. They are weaned in about six or seven weeks. They reach the adult length in about eight weeks and adult weight in about twelve to fifteen weeks.

The feeding habits of this species have not been studied, but mouse remains were found in one den, and the only two that Hamilton has examined

Figure 7.19. Distribution of the least weasel, *Mustela nivalis.*

contained the remains of deer mice. Two were examined by Whitaker from Indiana; one contained only *Microtus pennsylvanicus*; the other contained only vegetation, including grasses and mast.

Long-Tailed Weasel. *Mustela frenata* Lichtenstein

DESCRIPTION.   The long-tailed weasel is a long-bodied, short-legged carnivore; the male is nearly as large as a small mink. The head is short, the snout relatively blunt, the rounded ears set low; the tail is long and bushy, very noticeable in the live animal. Summer pelage is rich dark brown above; underparts, including the upper lips, white to deep yellow, the breast and belly often with pale brown spots; tail slightly darker than back, the terminal third or quarter black (Figure 7.16). A semiannual moult occurs in late fall and late winter or early spring. In the northern part of its range (New York to Wisconsin) a large proportion of the weasels become white, often stained with yellow, the tail tip alone remaining black. In areas where the winter coat is not white, the color is a pale shade of brown, much lighter than in summer. Males are much larger than the females. The skulls of males are 47–50 mm; of females, 40–42 mm. Average measurements of twenty adult males from New York are: total length, 405 (374–447) mm; tail, 135 (124–157) mm; hind foot, 44.5 (42–50) mm. Thirteen New York females average: total length, 325 (306–362) mm; tail, 107.5 (95–117) mm; hind foot, 37 (35–41) mm. Weight of males was 200–270 grams; of females, 71–126 grams.

DISTRIBUTION.   The long-tailed weasel occurs throughout the eastern United States (Figure 7.20).

HABITS.   In the north, this weasel appears to prefer more open country than the smaller short-tailed weasel; in the south it inhabits the dense hammocks and fringes of cypress swamps. It is found in the sparsely wooded second growth and in the thickets of low-growing shrubs and along watercourses, and it even invades extensive marshes, wherever there is a promise of abundant small life upon which to feed.

The weasel is a fearless little beast, hunting among the stone walls and cut-over brush land for its food. If an observer remains quiet, it will run about his feet, and its insatiable curiosity has often proved its undoing.

As cold weather approaches, the weasel sheds its brown summer coat for a white one. This change occurs during November, and usually occupies from three to four weeks. In northern New York and the higher parts of New England, nearly all the weasels become white; considerably less than half become white in Pennsylvania. Farther south practically none become white. The vernal change to brown usually occurs in March.

Figure 7.20. Distribution of the long-tailed weasel, *Mustela frenata.*

Weasels in search of food often cover several miles in a night, but these journeys may not encompass an area greater than a dozen or two acres.

Weasels make their home in a shallow earthen burrow, underneath a large stump, or in the bank of a gully. At the end of a tortuous tunnel a large nest is built, often composed largely of the remains of victims. While weasels climb with agility, they never seem to make their homes in trees.

The chief food of the weasel consists of small rodents; field mice, because of their abundance, provide a major share of the diet. In addition, the larger weasels successfully stalk and overcome cottontail rabbits, cotton rats, wood rats, chipmunks, shrews, small birds, and snakes, and do not disdain insects and even earthworms. They seldom if ever are attracted to vegetable matter. Weasels do not habitually suck blood, for stomach analysis has repeatedly demonstrated that these little carnivores eat flesh, bones, feathers, and fur. When attacking an animal nearly the size of itself, the weasel rushes in, grasps its unfortunate victim by the base of the skull and partially curls its serpentine body about the hapless prey, grasping it firmly with its forelimbs. A study of the stomach contents of more than 200 long-tailed weasels shows that more than 96 percent of the food consisted of small mammals, chiefly mice.

Weasels presumably mate in July, but the embryos pass through a long quiescent period of delayed development, the four to eight young not being born until mid-April. Obviously there can be but a single litter a year. The young are not weaned until they are five or six weeks old. Both parents bring food to the growing youngsters, which remain with the parents until midsummer.

Weasels have many enemies. Foxes, wildcats, the larger hawks and owls, and even the house cat are all known to catch and eat them.

Occasionally weasels visit the poultry yard, and with disastrous results. On these rare occasions, they may kill wantonly, destroying far more than they can eat at a time. On the whole, they are a distinct asset to the agriculturist, for their destruction of injurious rodents is of great benefit. In addition the larger weasels are profitable to the trapper, and many thousands of "ermine" pelts reach the fur markets each year. Large northern pelts are worth from thirty cents to a dollar, depending on market prices. Ermine pelts brought only twenty-five cents on the 1976 market.

Mink. *Mustela vison* Schreber

DESCRIPTION.   A weasel-like mammal with long lithe body, the mink has short legs, a long neck, and a comparatively short head which tapers to a rather pointed snout; the tail is long and bushy and the fur soft and lustrous, overlaid with longer glistening guard hairs; the toes of the hind feet are slightly webbed. Color is dark glossy brown, occasionally almost black, with white chin and variable white spots on throat; the skull is small, without sagittal crest; the anal glands are well developed. Males are larger than females.

Average measurements of eleven males from Quebec and the Adirondack Mountains, New York, are: total length, 535 (491–590) mm; tail, 174.5 (158–194) mm; hind foot, 60 (57–66) mm. Five adult females average: total length, 509 (481–597) mm; tail, 149 (144–155) mm; hind foot, 50 (47–54) mm. Males are 1.4–2.2 pounds (630–1000 g), females somewhat lighter (Figure 7.21). Individuals from most of the east are larger and paler than the preceding, and those from Wisconsin and upper Michigan are still larger. The richest color is attained in November; by midwinter the fur has commenced to fade and burn, owing to the activity of the animal in daylight.

Measurements of eleven adult males from New York, Maryland, and Georgia average: total length, 616 (547–652) mm; tail, 208 (180–238) mm; hind foot, 68.5 (54–81) mm. Five females average: total length, 507 mm; tail, 175 mm; hind foot, 56.5 mm. Large males from the swamps of western New York will weigh up to 3.5 pounds (1590 g), although the usual weight is 1.6–2 pounds (700–907 g). In Florida and coastal Georgia, where the species lives in the seemingly illimitable salt marshes, the mink is somewhat smaller. Measurements of ten adult males from there average: total length, 564 (526–615) mm; tail, 190 (179–203) mm; hind foot, 69 (61–72) mm. Females average 50 mm shorter in length. In Louisiana, the mink lacks the rich reddish brown coat of the northern form. Measurements of ten adult males from Louisiana average: total length, 563 mm; tail, 182.4 mm; hind foot, 70 mm.

DISTRIBUTION.   The mink occurs from Maine to Florida, westward to the Mississippi wherever watercourses provide a sufficiency of food and shelter (Figure 7.22).

HABITS.   The mink is found from the small streams of the spruce swamps to the tidal flats. It occupies diverse habitats but is seldom far from water.

E.L.Poole

Figure 7.21.   Mink, *Mustela vison.*

Figure 7.22. Distribution of the mink, *Mustela vison.*

During the winter, when the streams often freeze over, the mink will resort to the woods, occasionally occupying a rabbit burrow. It is not a rare animal, but one which is seldom seen except by the trapper.

The mink attains a larger size in inland marshes, possibly because of the better fare it can secure there. It makes a snug home under large trees which line the banks of streams, and whose large tangle of roots give it some measure of protection from its few enemies. Mink also occupy the lodges of muskrats, or natural cavities along the borders of rivers. The mink is a tireless wanderer, making extensive journeys, often cruising over a wide range and returning with marked regularity to the various parts of its range.

Mating takes place from January through March, but implantation is delayed. The three to six (sometimes eight) young are born about April, thirty to thirty-two days after implantation occurs. At birth they are the size of a cigarette, covered with fine, almost white hairs which scarcely conceal the pink flesh. The eyes open and the young are weaned at five weeks. Both parents assist in bringing food to the growing youngsters. The young follow the adults until well into the summer; then the family disperses. At this season the entire family may sometimes be seen hunting along the border of some quiet lake, but normally mink are nocturnal.

The mink relies upon fish, frogs, aquatic insects, snakes, small mammals, and such birds as it can secure. In extensive swamps where muskrats abound, the mink may feed largely on these big rodents, which are scarcely a match for their smaller enemy. Hamilton has found the winter nest of a mink composed almost entirely of the fur of muskrats.

Mink are provided with prominent anal glands which emit a powerful effluvium, particularly pronounced during the mating season.

The mink has few enemies. The larger owls, bobcat, and fox are known to take them on occasion, but their mortal enemy is man. Trappers take many thousands each winter, for the durable and lustrous pelt commands a good price. Trappers and hunters with trained dogs are largely responsible for the decrease in numbers of the mink over much of its range. A great number are taken in traps set for muskrats. Many were raised on fur farms in the 1930s and 1940s, about 200,000 being marketed annually, but relatively few are now raised for their pelts. In 1976, 40,000 wild northern mink pelts from New York averaged $15.

Badger. *Taxidea taxus* (Schreber)

DESCRIPTION.   The badger is a stout-bodied, short-tailed, short-legged weasel, easily recognized by its distinctive coloration and the large and heavy claws of the forefeet. The upperparts, including the tail, are clothed with long shaggy hairs, which are white or yellowish white at the base, pale brown to black in the middle, and prominently tipped with white, which gives a distinctly grizzled appearance to the animal. Nose, crown and neck are dark brown to blackish, the hairs on the neck are tipped with white; a large crescent in front of the ears is brown; the rest of the face and a slender stripe from nose to shoulders are white. Fur of underparts is short and whitish; legs are dark brown, feet black. Middle claws of forefeet are an inch or more long. Measurements of adult animals average: total length, 700–800 mm; tail, 120–155 mm; hind foot, 100–120 mm. Full-grown specimens weigh 15 to 25 pounds

(6.8–11.4 kg), the males being larger and heavier than the females (Figure 7.23).

DISTRIBUTION.   In the eastern United States, the badger ranges from northern and western Ohio westward through Michigan, northern Indiana, northern Illinois, and Wisconsin, where prairie or cutover land is found. L. E. Hicks has indicated that badgers have always occurred in Wood, Hency, and Fulton Counties in northwestern Ohio. They are found principally in the sandy areas of the old post-glacial lake beaches (Figure 7.24). At least five were trapped in the western Adirondack Mountains of New York in the 1930s, but these were unquestionably animals which had been recently introduced by man.

HABITS.   In the eastern part of their range, badgers frequent the open prairie country and flat rolling farm lands, shunning the woods and marshlands. There is increasing evidence that these animals are extending their range. Their expansion may be due to lumbering operations, which encourage the establishment of the rodent populations on which badgers are in large measure dependent.

The badger is admirably adapted for a fossorial life. Its long, stout foreclaws, heavy forelegs, and squat strong body enable it to dig with tireless energy and with amazing speed. Indeed, a badger can "dig itself in" in a very few minutes.

The bulky nest of dried grass is situated in an enlarged chamber at the end of a burrow which varies considerably in length and depth. Some have been

Figure 7.23. Badger, *Taxidea taxus.*

Figure 7.24. Distribution of the badger, *Taxidea taxus.*

noted that were six feet deep and thirty feet long, although other chambers are only a matter of six feet in length and one or two feet deep.

The animal is largely nocturnal, although it may occasionally be seen in the early morning or more particularly in the late afternoon and early evening, just as dusk is approaching. Inasmuch as it is usually a wary beast, the presence of the badger may go unnoticed for many years in a region where it is tolerably common. It is a strong and courageous fighter and more than a match for most dogs, for its loose skin affords little chance for a vital grip, and its strong teeth are capable of inflicting deep wounds.

The badger may retire to its underground chamber for several weeks or perhaps months during the severest part of the winter, but it does not truly hibernate and stirs about during mild spells.

The little evidence available suggests that the badger mates during the late summer; after a long period of interrupted development, the young, numbering one to five, are born during late March or April. They remain in the den until of good size. Little is known of the breeding habits, and some young naturalist located in an area where these animals are common could make a real contribution by a careful study of this subject.

The badger is the archenemy of ground squirrels, and its workings in a colony of spermophiles have all the appearance of an aerial blitzkrieg. No spermophile burrow is too long or tortuous for the badger to excavate, and each individual accounts for scores of these rodents every year. In addition to spermophiles, gophers, and other small rodents, it feeds upon insects, ground-nesting birds, snails, and other small animal life.

Other than man, the badger has few enemies. Its coarse but handsome fur was earlier used principally for trimmings, and for "pointing" more valuable furs, principally fox. Badger fur was likewise used in the manufacture of shaving brushes. Like many of the other larger species of mammals, however, the badger has become increasingly uncommon in many areas in the east, and is now generally protected. Many badgers are still killed by trappers and farmers, even though they are important in the destruction of noxious rodents.

Eastern Spotted Skunk. *Spilogale putorius* (Linnaeus)

DESCRIPTION.   A strikingly marked mustelid of moderate size, the spotted skunk is generally black in color, striped with white, the dorsal white stripes broken into patches on the hind quarters. There are four uninterrupted white dorsal stripes to the middle of the back, the center ones somewhat narrower, these stripes breaking into patches caudad; white patch on forehead and one in front of ear small; the *long* tail broadly tipped on its terminal fourth or third with white. Average measurements of five adults of both sexes from Alabama (males average slightly larger than females) are: total length, 540 mm; tail, 201 mm; hind foot, 48 mm; weight, 2–4 pounds (900–1800 g). Individuals from peninsular Louisiana and Florida are smaller. Adult males from Louisiana averaged a total length of 430–530 mm; tail 160–210 mm; hind foot 45–52 mm. Five adults of both sexes from Florida averaged: total length 332 mm; tail 113 mm; hind foot, 37 mm. Spotted skunks from Louisiana have the amount of white reduced whereas those from Florida have it increased (Figure 7.25).

Figure 7.25. Eastern spotted skunk, *Spilogale putorius.* Drawing by Lloyd Sandford.

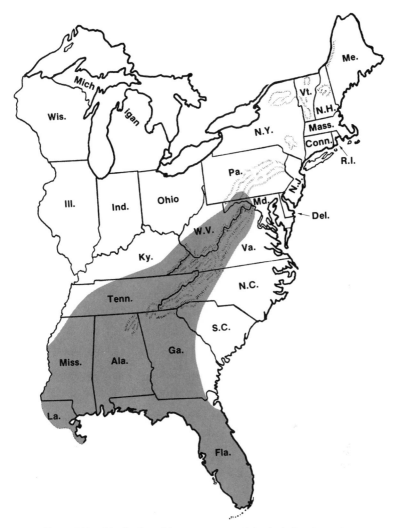

Figure 7.26. Distribution of the eastern spotted skunk, *Spilogal putorius.*

DISTRIBUTION. The spotted skunk ranges north in the Alleghenies to extreme southern Pennsylvania, thence southward in the western parts of the south Atlantic states to Florida, and Louisiana. There is an unreliable sight record from southern Illinois, but so far as we are aware, no specimens have been secured from this state (Figure 7.26). The species has been found in Indiana cave deposits, but there is no evidence that the species still exists in that state.

HABITS.   The beautiful little spotted skunk is common in much of the south, inhabiting waste land, cultivated fields, and the vicinity of farms; it seems to avoid heavy timber and wet lands. It often makes its home under the foundations of an outbuilding, or adopts a deserted woodchuck burrow or even the tunnel of a gopher turtle.

Spotted skunks are playful creatures, often romping with others of their kind like kittens. They are not easily intimidated by man, and, if suddenly alarmed, may actually advance upon the supposed adversary, occasionally walking for a few paces on their forefeet alone. The tail is usually held in an upright plume, always ready to be thrown over the back so that the rank effluvium may be discharged from the anal glands at the attack of an enemy.

In Texas, the spotted skunk is often referred to as the "hydrophobia skunk," as it appears to have a penchant for biting the sleeping cowherder about the face or other exposed parts. It has even been known to enter an adobe dwelling and bite a sleeping child. Several instances have been recorded where these bites resulted in rabies.

Mating occurs in the late winter, the females producing four or five young in the early spring. Little is specifically known of the breeding habits, but gestation is probably nearly sixty days. By twenty-one days the young resemble the adults in pattern. The eyes open at thirty-two days, and musk can be emitted at forty-six days.

Spilogale, like its larger cousin, the striped skunk, is an omnivorous feeder. During the winter and spring, it feeds largely upon rabbits, mice, and other small mammals; during the summer and fall season of plenty it fattens on fruits, insects, and birds. Lizards, small snakes, and offal are not disdained, and the spotted skunk has been rightly accused of stealing eggs and killing young chicks. It roots in the peanut fields for the luscious oily nuts, and feeds on fallen persimmons and various fruits in season.

The pelts of these little mustelids are much in demand, for they make up into beautiful jackets. The price paid the trapper for the hides seldom exceeds a dollar, but in spite of this many thousands reach the market annually.

Striped Skunk. *Mephitis mephitis* (Schreber)

DESCRIPTION.   So well known as to need little description, the striped skunk is a medium-sized, robust animal, characterized by the long black fur with two prominent white dorsal stripes and a long bushy tail; the feet semi-plantigrade, with large individual foot pads; forefeet with prominent claws adapted for digging; anal glands well developed; skull long and relatively narrow between the orbits; dorsal profile of skull characteristically angulate; broadly spreading zygomata, audital bullae little inflated and constricted; last upper molar quadrangular, very large, with an outer cutting ridge; palate ending either squarely or with a notch or spine. Males are larger than females (Figure 7.27).

The species reaches its largest size in northern Michigan where the basilar length of the skull in adult males is 72–75 mm long. In northeastern populations the size is large to medium; tail usually black, the tip white. The amount of white is variable: the dorsal stripes are often broad, nearly meeting on the back; but frequently the stripes are absent, the white restricted to a dash on the nape. Basilar length of male skulls is 60–66 mm. Measurements of eight New York adults average: total length, 603 mm; tail, 246 mm; hind foot, 65 mm. Skunks of the deep south are of medium size; tail very long, prominently marked with white on the sides, with a long white pencil; dorsal white stripes (sometimes absent) usually very broad; skull highly arched over frontals; rostrum very broad. Measurements of fourteen adults from Florida, the Great Smoky Mountains, and Georgia average: total length, 678 (598–719) mm; tail, 293 (215–351) mm; hind foot, 70 (64–76) mm. Individuals from northern Illinois are the smallest of the eastern striped skunks; white stripes narrow, usually terminating in middle of back; tail short, totally black, or with a white stripe; skull small and narrow without palatal spine; audital bullae more pronounced than in others. Measurements of three Illinois adults average: total length, 631 mm; tail, 196 mm; hind foot, 66 mm; weight, 4–10 pounds (1.8–4.5 kg).

DISTRIBUTION.   The striped skunk is widely distributed over the eastern United States, from Maine and Wisconsin south to Florida and Louisiana (Figure 7.28).

Figure 7.27. Striped skunk, *Mephitis mephitis.* Drawing by Lloyd Sandford.

Figure 7.28. Distribution of the striped skunk, *Mephitis mephitis.*

HABITS.   The skunk is well known, not on account of its handsome coat and gentle ways but because of its malodorous qualities. It is most abundant in agricultural areas, where the deep forest gives way to open lands and farm crops. Here it finds an abundance of food and the company of man, which it appears to welcome in spite of persecution.

A woodchuck burrow, the sanctuary of an outbuilding, or a burrow dug by the skunk itself serves as a retreat. In an enlarged chamber a warm nest of dried grass is made and here several individuals may pass the day, awakening at dusk to forage for food. Nightfall transforms this pretty animal into an active nimble little creature; it runs about in the moonlight, flipping stones deftly to one side to snare unsuspecting crickets or sleeping beetles.

As cold weather approaches, skunks become less active. A time comes when food is scarce, ice and snow cover the ground, and the skunk prefers to sleep for long days in its snug nest. It is not the dormant inactivity of hibernation, for a warm spell tempts the skunk from its lair, and the telltale tracks leave an apparently aimless trail through the snow-covered fields and woods. As many as fifteen skunks have been taken from a single large burrow. Usually the female and the young of the year are the soundest sleepers.

In northern climates, skunks become very active in mid-February. The mating urge is on them, and the woods are covered with tracks of wandering males searching a mate. After a fifty-one-day gestation, four to seven young are born in a blind, helpless condition, but with the striped pattern evident. Following a six- to seven-week period of nursing, the mother escorts her young from the home chamber on a warm June night, and they are introduced to hunting. A mother with her young following in serpentine fashion is a pleasant sight.

Skunks are not choosy of their food, and often make a nuisance of themselves about camps by raiding the garbage can. In the spring they eat quantities of field mice, May beetles, and their grubs. Later they haunt the wild raspberry and blackberry thickets, filling themselves with the fruit until they can hardly waddle. As fall approaches, wild cherries, crickets, and grasshoppers are their mainstay, and these tidbits all add to the substantial layer of fat which the skunk has been acquiring since midsummer. It is thus able to sustain life during the rigorous winter when food is scarce. This fat is utilized during the colder months, for skunks may lose from 15 to 30 percent of their weight or even more from November to March.

The skunk's chief claim to fame rests on its powerful effluvium, housed in the large anal glands. The glands are easily made useless by snipping the nipples just within the vent. Adhesive inflammation follows that occludes the ducts at this point, effectively sealing the glands. The glands are used only in times of great stress, but the warning of the uplifted tail is often sufficient to intimidate all but the most insistent foe. The mephitic odor does not appear to deter predation by the horned owl, bobcat, or fox.

The thick fur of the skunk makes an attractive garment, and many thousands are trapped each year. In 1976, 40,000 New York pelts averaged only $2, contrasted to $12 for a good pelt in 1928. In addition to the mortality caused by the traffic in fur, thousands are destroyed on the roads every year. The skunk is normally deliberate in all its movements and apparently has not learned the hazard of the highway. Hamilton once counted twenty-seven dead animals on a thirty-four-mile stretch of paved road in New York during

October, but this was before rabies took its toll and decimated the northeastern population. The striped skunk has now supplanted the dog as the United States species most often contracting rabies.

River Otter. *Lutra canadensis* (Schreber)

DESCRIPTION.   A large aquatic member of the weasel tribe, the otter has a broad, flattened head, small eyes, small ears, long stout neck, and heavy tapering tail, which serve as good field marks. The short legs and webbed toes, hairy soles, and thick dense pelage are characteristic. Color above is a rich glossy brown, occasionally nearly black; underparts much lighter, the lips and cheeks, chin, and throat a very pale brown. The skull is flattened, rostrum short, cerebral portion swollen backward and outward; audital bullae much flattened; upper carnassial tooth with tricuspid blade and large inner lobe. Total length is 900–1100 mm; tail, 300–400 mm; hind foot, 100 mm. Weight is 12 to 18 pounds (5.4–8.2 kg), seldom more (Figure 7.29). Otters along the coast from New Jersey to South Carolina are paler in color than the typical form. Those from the southern parts of the range of the species are larger. Individuals from Florida have a longer tail and less black; total length is 1100 to 1300 mm.

In his taxonomic and systematic revision of the genus *Lutra,* C. G. van Zyll de Jong (1972) concluded that the New World otters are generically distinct from *Lutra* of the Old World and for the former he resurrected the name *Lontra* (Gray). I. I. Sokolov (1973), on the basis of a morpho-anatomical analysis of recent and extinct forms of the subfamily Lutrinae, concluded that *canadensis* should be retained in the genus *Lutra.*

DISTRIBUTION.   Originally, this species occurred throughout the eastern United States from Maine and the region of the Great Lakes to the Gulf of Mexico (Figure 7.30). However, the species has been extirpated in many parts of its range. The last known record of the otter in Indiana, for example, was in 1942.

HABITS.   The otter is a shy and retiring animal, often occupying regions close to man; but owing to its secretive habits it is seldom seen. It frequents the lakes and large rivers of the northern wilderness forests, the lonely bayous and cypress lined rivers of the south, and occurs in some numbers in the extensive marshes of the Gulf. Within recent years it appears to have increased and actually become a nuisance in Massachusetts, where it invades the fish hatcheries and prized trout streams. The otter appears to favor fresh water in the north, although it is found in brackish streams in the south.

It is a graceful swimmer, moving in an undulating course, or swimming with only the head above the surface. It is an expert diver and can swim long

Figure 7.29.   River otter, *Lutra canadensis.*

distances beneath the surface, traveling for a quarter-mile without coming up for air if the need arises. It need not travel far in the summer, for food is easily secured and life is easy. But where the winters are long and cold, it must make long overland journeys to visit the infrequent pools kept open by cascading falls, for only here can it secure the fish which maintain it at this season. In spite of its short legs, the otter moves readily on land, and on the ice it expedites travel by sliding.

The otter is famous for its slides, which in the north are made on steep slopes coated with ice and snow; in the south a clay bank serves the same purpose. On these the otter slides to the pool below, its forelegs tucked under it. The slides soon become slick and several otters may join in the sport of tobogganing.

The time of mating is not definitely known. Delayed implantation occurs. The young are usually born from mid-April to early May. They normally number two to three, and are fairly well developed at birth. The natal chamber is usually a burrow situated beneath the sprawling roots of some tree bordering a watercourse. When the young are weaned, they are taught by the mother to swim, for they must learn as any other animal. The mother is said to take the youngsters on her back at first. When they have lost fear of the water, the mother dives, returning to support them until the young are at last accomplished swimmers.

Much has been written on the supposed destructiveness of otter to trout, and without question they successfully pursue and capture these fish. Hamil-

Figure 7.30. Distribution of the river otter, *Lutra canadensis*.

ton has seen rocks in "otter" streams literally plastered with the scales of these fish. Suckers and crayfish are important foods, with other shellfish, frogs, and small animal life generally. A study of 141 Adirondack otters by Hamilton revealed that fish occurred in 70 percent of stomachs, with 5 percent of these being trout. Crayfish, frogs and tadpoles, and aquatic insects comprised other important foods. The otter is said to kill muskrats, but there seems to be no direct evidence for this assumption.

The pelt of the otter is thick and lustrous. It is the most durable of native American furs. Prime 1976 pelts brought $60. Many pelts are exported annually. The pelts of both the river otter and sea otter were the favorite furs of the Chinese mandarins and the Russian nobility.

Mountain Lion. *Felis concolor* Linnaeus

DESCRIPTION. The mountain lion (also called panther or cougar) is a large, pale brown, unspotted cat with relatively small round head and long, dark-tipped tail; limbs stout and thick; color dun to reddish brown above, shading into dull white on the under parts; inner ear, lower cheeks, chin, and lips white; a dark spot at the base of the whiskers (Figure 7.31). Total length extends to 7 feet. Five males averaged 2,257 mm (89 inches); four females 1,984 mm (78 inches). Tail and hind foot of type specimen, an old adult male,

Figure 7.31. A large mountain lion, *Felis concolor,* shot in Lee County, Florida, in 1940.

were 760 and 280 mm respectively, whereas corresponding measurements for an old adult female from the same locality were 670 and 271 mm. Young are yellowish-brown with irregular brownish-black spots which are usually lost when the kittens are about six months old. There are many reports of black panthers throughout the range of the species, but to date the only authenticated case of a black panther is from Brazil.

A large specimen shot a mile from Bonita Springs, Lee County, Florida, on October 10, 1939, weighed 145 pounds (66 kg) after considerable loss of blood. A very large individual killed near Estero, Lee County, Florida, on November 20, 1939, was estimated to weigh less than 200 (90 kg) pounds. Hamilton saw a freshly killed female on April 3, 1941, at Fort Myers, Florida, the weight of which he estimated at 150 pounds (68 kg).

DISTRIBUTION.   This species once occurred throughout the eastern United States, but the wild, impenetrable swamps and hammocks of southern Florida are the last real stronghold of the mountain lion east of the Mississippi (Figure 7.32), and the state of Florida has imposed total protection. In spite of considerable hunting and persecution, it appears probable that the animals will persist for some years to come in the unsettled Everglades, especially in view of present conservation efforts. There are records of specimens being shot in 1939, 1940, and 1941 at Naples, Estero, Bonita Springs, and Immokalee, and James N. Layne has accumulated many recent records for Florida. Layne estimates the Florida population at perhaps 100 to 200 individuals.

Repeated press accounts of the presence of these large cats in northeastern United States occur, but most such stories lack authenticity. The Adirondacks were the last stronghold of the mountain lion in the northeast, and the latest record of one from this wilderness was in 1903. A mountain lion was trapped at Little St. John Lake, Maine, in 1938 and was mounted (Wright 1961). A cougar was reportedly killed on the road near Georgetown, South Carolina, in 1942 or 1943. One was shot in Tuscaloosa County, Alabama, in 1956, plaster casts of cougar tracks were made in Clarke County, Alabama, in 1961, and a reliable sight observation was made in the same county in 1966. One is listed as a probable roadkill from the Massachusetts Turnpike in 1960, and one was shot and preserved at Edinbora, Pennsylvania, in 1967 (Doutt, Heppenstall, and Guilday 1967). W. C. Dilger observed unmistakable snow tracks and tail markings of a mountain lion at a deer kill near Ithaca, N.Y., in January 1976. There are a number of recent records from New Brunswick, and there may be a remnant population in Maine. Some believe the cougar could still survive in such wilderness areas as the Great Smoky and Blue Ridge Mountains. However, throughout the eastern United States, mountain lion and wolf reports run rampant, even in Indiana, where neither species has existed for many decades. This makes it difficult to rely on records not supported by specimens.

Figure 7.32. Present distribution of the mountain lion, *Felis concolor,* in the eastern United States.

HABITS. This large, shy cat is rarely seen, and its occurrence is usually noted only by the spoor. It may be found in many different environments, showing little habitat preference other than for generally wooded situations over extensive open lands. It often uses swamps or wooded water courses. It seldom is abroad during the day, preferring to sleep in some secluded area and journey forth at nightfall, when the deer and other large animals upon which it feeds commence to move above. The mountain lion is a wide ranger, moving perhaps twenty or thirty miles on its hunting trips. Its habitat is very different from that of the western cats, the habits of which are much better known. It may breed throughout the year in Florida, producing one to three young per litter (averaging about two). Elsewhere, up to six young have been recorded in a litter. On April 3, 1941, Hamilton was privileged to be present when a cowhand brought into Fort Myers, Florida, a large female which he had shot earlier in the day near Immokalee. He exhibited its two kittens, estimated to be about a month old. The gestation period is three months. The development of the kittens is not unlike that of the domestic tabby. The eyes of the kits open at eight or nine days, the incisors and canines cut the gums in somewhat less than three weeks, and weaning occurs when the kits are about three months old. The young number from one to three, two or three being usual.

Deer seem to provide much food for the lion over most of its range, but many other foods are eaten. Domestic livestock are occasionally taken, and also there are records for raccoons, skunks, foxes, rabbits, armadillos, birds, turtles, turtle and bird eggs, alligators, snakes, and insects. In July 1940, Dwight Dyess was riding on the trail of a lion, in the hammocks about seventeen miles east of Fort Myers, Florida, when he came upon the still-warm body of a half-eaten pig. It was a sow that had been pulled back into the palmettos; the body was covered with several branches. Dyess has also found that the fetus of a gravid doe, killed by a lion, had been removed and eaten. A mare with colt will occasionally be killed and only the fetus eaten. The lion kills its larger prey, such as a deer, heifer, or colt, by rushing and throwing the weight of its entire body against the victim. It usually grabs the unfortunate animal in the throat, back of the neck, or the breast. After making a meal, the lion fequently covers its prey, returning when it is again hungry.

The ability of the lion to scream has been disputed by many, but it is now generally recognized that these animals are capable of emitting unearthly wails. Charles B. Cory, an eminent naturalist who spent much time in Florida in the latter part of the past century, remarks that the "cry of the old panther somewhat resembles the screech of a parrot, but is much louder." Ned Hollister (1911) describes the cry of the Louisiana panther as "a long-drawn-out, shrill trill, weird and startling. It commences low on the scale, gradually ascends, increasing in volume, and then lowers at the end."

Key to Cats of the Genus Lynx

A. Tip of tail black above and below; one large, one small foramen at posterior inner corner of tympanic bulla . . . .*Lynx lynx*

AA. Tip of tail black above only; these two foramina confluent, leaving one opening . . . .*Lynx rufus*

Lynx. *Lynx lynx* (Linnaeus)

DESCRIPTION. The lynx is a short-tailed, medium-sized cat with large feet, prominent ear tufts and ruff of long brown-tipped hairs on sides of head; tail tip wholly black. Its color is light gray, sprinkled with pale brown, often a pinkish buff hue in tone; no prominent black or brown spots on belly or flanks. Total length is 825–950 mm; tail, 100–125 mm; hind foot, 215–250 mm. A large male from Ripton, Vermont, weighed 35 (16 kg) pounds. Forty pounds probably approaches the weight limit (Figure 7.33).

DISTRIBUTION. It is difficult to define accurately the present range of the lynx in eastern United States, as few specimens are available, and trappers often confuse this species with the bobcat. It is not common in northern Maine; a specimen was secured at Ripton, Vermont, on November 5, 1937. It is scarce in New Hampshire but ten, nine, and five bounties were reportedly paid for lynx in the periods 1951–55, 1956–60, and 1961–65. The bounty was removed in 1965. The lynx is scarce in the Adirondacks. A pair was killed in Willseyville, New York, during the spring of 1907. A few persist in the Upper Peninsula of Michigan and there are recent records for Iron, Washburn, Sauk, and Taylor Counties, Wisconsin, and for Marquette, Michigan. Elsewhere in the east it has disappeared during the twentieth century (Figure 7.34).

Figure 7.33. Lynx, *Lynx lynx.*

Figure 7.34. Distribution of the lynx, *Lynx lynx.*

HABITS. The dense coniferous forests of northern New England are the home of the remaining lynx in northeastern United States. It dwells only in the shaded spruce forests where subdued light alone reaches the forest floor. Here in the snow-covered dense timber and swampy thickets it travels on big, thick-furred pads, ever on the alert for the ghostly white hares which sustain it in the frigid winter months.

Removed from its habitat of windfalls and logs, moss-covered boulders and the uneven floor of the forest, the lynx is surprisingly slow. In open country it moves in a series of gallops, and is soon overtaken by a dog.

The lynx once ranged as far south as Indiana and Pennsylvania, but many years have passed since its disappearance from these states. The lynx is an adept climber, and it is thought that it drops from the low limbs of trees onto the back of its unsuspecting prey. An able swimmer, the lynx is known to cross sizable lakes.

Little is known of the home life of the lynx. It is said to mate in the late winter, and after a probable gestation of two months the one to four young are born in some dense windfall or rocky ledge. Their eyes open when they are ten days old, but the kittens are streaked and spotted at birth and remain so colored for several months. As with all wild Felidae, a single litter is produced each year, and the young first breed in the winter following their birth.

These big cats subsist largely upon the varying hare; when these are scarce, famine stalks the lynx world. In addition to hares, mice, squirrels, skunks, and even such large prey as foxes are killed. It is not difficult for a lynx to overcome a large buck, particularly in periods of deep snow, for the big feet of the lynx act as effective snowshoes whereas the deer flounders helplessly about until exhausted.

Other than the northern trapper, the lynx has little to fear. Many are taken, for the pelts are dense and lustrous. The fur is used principally in scarves and coat trimmings.

Bobcat. *Lynx rufus* (Schreber)

DESCRIPTION.   The bobcat is a medium-sized, short-tailed cat of reddish brown tone, streaked with black; underparts white with black spots; back of ears black with a central white spot. The black ear tufts are usually less than an inch long and often wanting. The fur is relatively short and sleek, contrasting sharply with the lynx. The color of the tail is distinctive; the tip is black above and white below, with three or four poorly defined brownish black bars on the distal part (Figure 7.35). In winter the fur is longer and fuller, but the feet are never padded as are the great "snowshoes" of the lynx. Fourteen adult Vermont bobcats measured by wardens averaged: total length, 988 (850–1,200) mm; tail, 127.5 (95–150) mm; hind foot, 158 mm. Remington Kellogg (1937) gives the measurements of nineteen adult bobcats from West Virginia. These averaged: total length, 834 mm; tail, 145 mm; hind foot, 165.5 mm. Seventy-four Vermont specimens whose weights Hamilton recorded averaged 15.25 pounds (6.8 kg). The largest, an old male, was 36 pounds (16 kg); seven weighed 30 pounds (13.6 kg) or more. Young cats weigh about 10 pounds (4.5 kg) in the fall of their first year. Southeastern bobcats are large but lightly built, with very small feet. They are darker than typical *rufus* and differ in color pattern, being much spotted and having black waved streaks on the back. Lowery (1974) records a large specimen taken near Bains, Louisiana, which measured 44 inches from tip of nose to tail and weighed 45.25 pounds (22 kg).

Figure 7.35. Bobcat, *Lynx rufus.*

DISTRIBUTION.   The bobcat is widely distributed from Maine to Florida westward to the Mississippi and often occupies wild areas close to large cities. In New Hampshire, bounties were paid on 1699, 1750, 1908, and 386 bobcats in the periods 1951–55, 1956–60, 1961–65 and 1966–70, respectively. The species is doing very well through much of the south where it is extensively hunted, although hunters now quite often "tree" the animal and then allow it to live as a manner of protection. Bobcats are now absent from much of the midwest, as well as large areas of Maryland, Kentucky and Tennessee (Figure 7.36).

Figure 7.36.   Distribution of the bobcat, *Lynx rufus.*

HABITS. The bobcat is adaptable, changing its way of life as civilization modifies the environment. It does not depend on the deep forest, as the lynx does, but rather makes its home in the swamps, wooded districts, and mountainous regions, often close to farm lands. In its southernmost range, the bobcat is found in cypress swamps and pine woods, often near large cities. Hamilton has found the dens of these animals in the Ramapo Mountains less than thirty miles from New York City. The species can exist in low numbers in remote areas for extended periods without being detected. It was long thought to be extinct in Indiana, although there were persistent reports in several areas. However, in 1970, in a rugged area of Monroe County, Indiana, one was shot by a hunter.

The bobcat is a good climber, and will usually tree after being run a short time by dogs. It is more than a match for the average hound; like many of the large cats, it seems to be more annoyed than frightened by the baying of a hound.

The home is often a sizable thicket, stump, or rock ledge. Hamilton has found several dens in mountainous country, all of which were in rock recesses. The odor of these dens, and indeed of the cats themselves, is very strong.

The mating season commences in late February, at least in the north, and the two to four young are born after a gestation period of about fifty days. The young are well furred and spotted at birth, open their eyes on the ninth day, and are weaned when about two months old. They remain with the mother well into the summer but by fall, when they weigh 8 to 12 pounds, the youngsters are hunting by themselves.

The larger northern bobcats are sufficiently strong to overcome a full-grown deer. The chief food of the bobcat during the fall and winter months is snowshoe hares, squirrels, mice, muskrats, grouse, jays, and smaller creatures. Hamilton has found the remains of red and gray foxes in the stomachs of large cats; porcupine, mink, and muskrat are other items of fare. A few stomachs which he examined from the Smoky Mountains of North Carolina contained the remains of beetles, rabbits, and a small box turtle.

The unprecedented rise in prices for long-haired raw furs in recent years has spurred interest in trapping. New York bobcat formerly brought only a few dollars, but prime pelts brought $90 in 1976.

8

Artiodactyla

(Deer)

The even-toed hoofed mammals are characterized by the reduction or complete suppression of all but the third and fourth digits. These are equally developed with the terminal phalanges flattened and provided with hoofs. The Artiodactyla are animals of large size, with broad, crowned, and ridged grinding molars, adapted for cutting browse and grasses. Except for Australia, they are world wide in distribution. Only one family, the Cervidae, occurs in eastern United States. This group is characterized by bony outgrowths from the frontal bones, called antlers, in the males (also in the female caribou) and the absence of incisors in the upper jaw.

White-Tailed Deer. *Odocoileus virginianus* (Boddaert)

DESCRIPTION.   The white-tailed deer differs from other members of the genus in its antlers, the main beam being directed forward and bearing the several tines behind. No brow tine is present. The long, rather bushy white tail and the small metatarsal gland margined with white hairs also serve to distinguish this species.

The seasonal color and character of the pelage vary markedly. In summer the coat is bright tan, darker along the middorsum, paler on the face, throat and breast. A prominent nose band, the orbital region, the upper throat, the inside of the ears, legs and belly are pure white. Each side of the chin bears a black spot.

In the fall the red summer coat is replaced by a dense, somewhat brittle "blue" coat, best described as a ticked brown. The black chin patch is less sharply defined, although the white hair is still sharply demarked from the gray portions.

The young are reddish yellow, spotted with white. These spots are lost by late summer. Weights vary considerably. A large buck from the Carolinas or New Jersey will weigh 200 pounds (91 kg); females average from 90 to 120 pounds (41–54 kg) (Figures 8.1 and 8.2).

Deer from extreme southern Florida are small, mature bucks seldom exceeding 150 pounds (67 kg) and the average weight considerably less. Deer from the Everglades are medium-sized or rather large; pelage very short; upperparts in summer pelage a dark rufescent shade near hazel; ears and hind feet rather short; antlers narrowly spreading; rostrum slender. Individuals from Louisiana are somewhat smaller than northern deer and have a massive dentition and pale drab color, the body liberally sprinkled with black-tipped hairs which form a distinct medio-dorsal line extending backward from the crown.

DISTRIBUTION. The white-tailed deer occurs throughout the eastern United States (Figure 8.3).

Figure 8.1. White-tailed deer, *Odocoileus virginianus,* head of buck. Photo by William Royce.

Figure 8.2. White-tailed deer, *Odocoileus virginianus,* doe and fawn.
Photo courtesy New Hampshire Fish and Game Department.

HABITS.   The white-tailed deer, once all but exterminated in the north-eastern and midwestern states, is now a common animal of the farmlands and open woods there. It occurs in the open glades, swamps, and valleys through-out the northern states, in the swamps and thickets of the south, on the coastal islands of the south Atlantic states, in the dense Florida hammocks, and even on the Keys. A surprisingly small thicket will conceal these creatures. One may pass within ten feet of a crafty buck and never see it; the head is pressed low to the ground and the great antlers blend admirably with the brush.

The deer is a gifted swimmer: it often takes to sizable lakes in order to escape the dogs and is known to swim from one to another of the small Florida Keys. The winter hair of the northern deer is so charged with air that it is almost impossible for the animal to sink if it should become exhausted while swimming.

Deer are active at all hours, although much less so during the day. They usually commence to feed with the approach of dusk and continue to move about during the night, usually bedding down with the approach of dawn.

The antlers, carried only by the males, are deciduous osseous structures; they are the most rapid-growing membrane bones of mammals. The antler grows from a pedicle of the frontal bone and is covered with a true skin, the velvet, during its period of growth. When full size is attained, the velvet is scraped off against small trees. Evidence of this is often seen where white-tails are common. The first antlers, usually consisting of a single spike, are obtained in the second year. Growth of the antler is rapid, full size being obtained in four months or less. After the mature antlers have been carried for several months, they become loosened and fall off. In the north the antlers are shed from late December to February, but in south Florida the bucks drop their antlers earlier. A buck in Lee County, Florida, during late March, had just begun to grow new antlers. It is commonly thought that the age of a buck can be determined by the number of tines, but this is not so. Nutritional

Figure 8.3.   Distribution of the white-tailed deer, *Odocoileus virginianus.*

factors greatly influence that number. Wildlife biologists have worked out fairly accurate aging techniques based on tooth replacement and wear.

At the mating season the buck, with polished antlers and great swollen neck, searches the woods for a doe, and savage battles are fought for the favor of a mate. Not infrequently the antlers of two bucks become inextricably locked, and after a desperate but futile battle, they perish. These deer are probably polygamous, but there is some reason to suspect that a buck will mate with only one doe. It is probably the least polygamous of all deer. In the northern states mating occurs in November and the young are born in late May or June, after a gestation of seven months. In Florida and the Gulf states summer is the time of mating; the young come in the period from January to March. About half of the northern does mate and conceive in their first year. A young doe usually has a single fawn, but twins are usual in a mature female, and occasionally three are born. The spotted fawns are concealed in some dense thicket, blending so well with the dappled shadows and standing so perfectly motionless that they are seldom seen. The fawns remain with the mother well into the fall or winter.

The deer is a browser but deigns to graze; indeed much of its summer food is of the herbs which border the woodland or the choice lily pads and pond plants which are easily obtained. With the ripening of acorns, beechnuts, and other mast, the deer fatten on these nutritious foods, so they are fat and sleek when winter arrives. Now their food must be browse, and they feed on the tender buds and twigs of maples, birch, viburnums, blueberries, white cedar, and many conifers. Many other plants have been recorded as food of the white-tail.

Winter is the time of hardship for the northern white-tail. Heavy snows and drifts cover the woods, and while the deer will herd to some extent, opening paths or yards in the brushy swamps, many perish from exhaustion and starvation. In early days the wolf was probably the greatest enemy of the eastern deer. Foxes will kill the fawns, and bobcats manage even the adults during the winter season, when the deer are partially helpless. Feral dogs presently are important predators of the white-tail.

The deer is the most important big game animal in the United States. In the northeast, between 150,000 and 300,000 deer are killed annually. In 1938, Pennsylvania hunters killed 127,000 deer, chiefly does; 40,000 bucks were killed in Michigan in the same year; in 1936, 30,000 were shot in Wisconsin. Maine, New Hampshire, Vermont, Massachusetts, New York, the Carolinas, and Louisiana are other states in which large numbers are killed. In the 1974 season, deer hunters in New York State killed 103,303 deer, a record take for the state. Some states where deer have been reintroduced after earlier extinctions, such as Indiana, now have ample herds and good hunting.

Moose. *Alces alces* (Linnaeus)

DESCRIPTION.   This great beast is the largest of all living deer, and can be mistaken for no other animal. The massive creature, with high humped shoulders, broad pendulous muzzle, long coarse black hair, and whitish legs is an impressive animal. The ears are large, and there is a prominent hairy dewlap on the throat. The bull has broad palmate antlers. An estimate based on dressed weight indicates that a large bull may exceed 1400 pounds (624 kg). Winter pelage is blackish brown, not noticeably paler beneath; the lower parts of the legs are light brownish gray, giving one the impression of stockings (Frontispiece).

DISTRIBUTION.   The moose formerly occurred over much of New England, the Adirondacks of New York, extreme northern Pennsylvania, and northern Michigan and Wisconsin. It is still common in northern Maine, and a few individuals are recorded from northern New Hampshire and Vermont. These may have moved in either from Maine or Quebec. Moose have been recorded in western Massachusetts since 1932, but they are probably escaped, semicaptive individuals from an estate in central Berkshire County. A few have been recorded from the Adirondacks of New York in the past thirty years. The moose is yet found in the upper peninsula of Michigan and occurs in some numbers on Isle Royale in Lake Superior (Figure 8.4).

HABITS.   This great deer frequents the spruce forests, swamps, and aspen thickets which border the northern lakes and rivers. It is a majestic beast, inspiring awe in one who first sees it. In spite of its huge bulk and tremendous palmate antlers, the moose is capable of trotting through heavy brush or the forest with amazing speed and stealth. It often haunts the waterways, particularly in the summer, both to seek relief from the hordes of flies and to secure the tender roots and stems of the water lilies.

Figure 8.4.   Distribution of the moose, *Alces alces.*

Moose lead a solitary life except in the winter, when the deep snows limit movement. Then a band forms, the combined efforts of which pack the snow down for considerable areas. Within this "yard" the moose move easily, searching out the tender twigs of many trees. Like the deer, they are browsers during the lean months, but the food habits undergo a change when the growing season returns. It is then that they repair to the shallow lakes seeking the submerged rhizomes, petioles, and leaves of the spatter-dock (*Nymphaea*). Hamilton watched a huge bull feeding in such a small northern pond on the slopes of Mount Katahdin, in northern Maine, during August. At times the bull would become quite submerged and remain completely out of sight for many seconds. Then the huge head with velvet-covered antlers would break the surface and the moose would contentedly chew its prize.

Moose are good swimmers, striking boldly across large lakes and swimming powerfully. Only an accomplished canoeist will be able to overtake a moose in the water.

The mating season commences in mid-September and lasts for a month. With the lust upon them, the huge bulls rush through the forest, ever alert for the grunting cows, or challenging rival bulls with their bellows. Titanic struggles are fought; the vanquished flees for his life. There is a difference of opinion as to whether the bulls mate with more than one cow, but the best opinion favors polygamy. Mating is usually over by mid-October. Following an eight-month gestation, the single or twin calves are born. These are paler, more reddish brown, than the parents, but unlike the deer fawns, the moose calf is never spotted. The young is weaned at about six months but remains with the mother for about a year.

Hunters call the bulls with crude birch bark horns, simulating the grunt of a cow. Often a stick rattled in an alder thicket will attract the great brutes in September, for at this season they too rattle their big antlers in the saplings to free them of the last vestiges of velvet.

Literature of Mammalogy

LITERATURE CITED

ADAMS, L. 1965. Biotelemetry. BioScience 15(2): 155–157.

ALLEN, A. A. 1921. Banding bats. Journal of Mammalogy 2(1): 53–57.

BAILEY, VERNON. 1924. Breeding, feeding and other life habits of meadow mice (*Microtus*). Journal of Agricultural Research 28: 523–535.

BARBOUR, THOMAS. 1936. *Eumops* in Florida. Journal of Mammalogy 17: 414.

BIRKENHOLZ, D. E. 1963. A study of the life history and ecology of the round-tailed muskrat (*Neofiber alleni* True) in north-central Florida. Ecological Monographs 33: 255–280.

BOWEN, W. W. 1968. Variation and evolution of Gulf Coast populations of beach mice, *Peromyscus polionotus*. Bulletin of the Florida State Museum 12.

BRADT, GLENN W. 1938. A study of beaver colonies in Michigan. Journal of Mammalogy 19: 139–162.

CLARK, D. R., RICHARD K. LAVAL, AND D. M. SWINEFORD. 1978. Dieldrin induced mortality in an endangered species, the gray bat (*Myotis grisescens*). Science 199(4335): 1357–1359.

CONNOR, PAUL C. 1960. The small mammals of Otsego and Schoharie Counties, New York. N.Y. State Museum and Science Service Bulletin 382.

CONSTANTINE, D. G. 1962. Rabies transmission by non-bite route. U.S. Public Health Reports 77: 287–289.

CORY, CHARLES B. 1896. Hunting and fishing in Florida. Estes and Lauriat, Boston.

COUES, ELLIOT. 1872. Key to North American birds. Vol. 1. Page Publishers, Boston.

DAVIS, W. B. 1938. A heavy concentration of *Cryptotis*. Journal of Mammalogy 19: 499–500.

DAVIS, WAYNE H. 1955. *Myotis subulatus leibii* in unusual situations. Journal of Mammalogy 36: 130.

DAVIS, WAYNE H., AND W. Z. LIDICKER, JR. 1956. Winter range of the red bat, *Lasiurus borealis*. Journal of Mammalogy 37: 280–281.

EADIE, W. ROBERT. 1939. A contribution to the biology of *Parascalops breweri*. Journal of Mammalogy 20: 150–173.

ENGELS, WILLIAM L. 1941. Distribution and habitat of *Sorex longirostris* in North Carolina. Journal of Mammalogy 22: 447

EVANS, F. C., AND R. HOLDENRIED. 1943. A population study of the Beechey ground squirrel in central California. Journal of Mammalogy 24: 231–260.

GIER, H. T. 1957. Coyotes in Kansas. Kansas Agricultural Station Bulletin 393.

GODFREY, GILLIAN K. 1954. Tracing field voles, *Microtus agrestis,* with a Geiger-Müller counter. Ecology 35: 5–11.

GODIN, A. J. 1977. Wild mammals of New England. The Johns Hopkins University Press, Baltimore, Md. 304 pp., illus.

GREEN, MORRIS M. 1930. A contribution to the mammalogy of the North Mountain region of Pennsylvania. Privately published, Ardmore, Pennsylvania.

HALL, JOHN, AND NIXON WILSON. 1966. Seasonal populations and movements of the gray bat in the Kentucky area. American Midland Naturalist 75: 317–324.

HAMILTON, W. J., JR. 1931. Habits of the star-nosed mole, *Condylura cristata.* Journal of Mammalogy 12: 345–355.

———. 1933. The insect food of the big brown bat. Journal of Mammalogy 14: 155–156.

———. 1934. The life history of the rufescent woodchuck, *Marmota monax rufescens.* Annals Carnegie Museum, Pittsburgh, Pa., 23: 85–178.

———. 1937. The biology of microtine cycles. Journal of Agricultural Research 54: 779–790.

———. 1940. The biology of the smoky shrew, *Sorex fumeus.* Zoologica, New York Zoological Society, 25: 473–491.

———. 1958. The life history and economic relations of the opossum in New York State. New York State College of Agriculture, Cornell University, Memoir 354.

———. 1974. Food habits of the coyote in New York State. New York Fish and Game Journal 21(2): 177–181.

HARPER, FRANCIS. 1927. Mammals of the Okefinokee Swamp region of Georgia. Proceedings of the Boston Society of Natural History 38: 191–396.

HARTMAN, CARL. 1920. Studies in the development of the opossum. Anatomical Record 19: 1–11.

HEIDT, GARY L. 1970. The least weasel, *Mustela nivalis* Linnaeus: The developmental biology in comparison with other North American *Mustela.* Publication of the Museum, Michigan State University Biological Series 4(7): 227–282.

HOLLISTER, NED. 1911. The Louisiana puma. Proceedings of the Biological Society of Washington 24: 175–178.

HOWELL, A. H. 1921. A biological survey of Alabama: The mammals. North American Fauna No. 45. Washington, D.C. 88 pp.

JACKSON, HARTLEY H. T. 1928. A taxonomic review of the American long-tailed shrews (genera *Sorex* and *Microsorex*). North American Fauna No. 51. Washington, D.C. 238 pp.

———. 1961. Mammals of Wisconsin. University of Wisconsin Press, Madison. 504 pp.

KELLOGG, REMINGTON. 1937. Annotated list of West Virginia mammals. Proceedings of the United States National Museum 84(3022): 443–479.

———. 1939. Annotated list of Tennessee mammals. Proceedings of the United States National Museum 86(3051): 245–303.

KENNICOTT, ROBERT. 1859. The quadrupeds of Illinois, injurious and beneficial to the farmer. U.S. Patent Office Report for 1858. Pp. 241–256.

KOMAREK, EDWIN V., AND ROY KOMAREK. 1938. Mammals of the Great Smoky Mountains. Bulletin Chicago Academy Science 5(6): 137–162.

LAYNE, JAMES N. 1954. The biology of the red squirrel, *Tamiasciurus hudsonicus loquax* (Bangs) in central New York. Ecological Monographs 24: 227–267.

LAYNE, JAMES N., AND W. J. HAMILTON, JR. 1954. The young of the woodland jumping mouse, *Napaeozapus insignis*. American Midland Naturalist 52: 242–247.

LAZARUS, A. B., AND F. P. ROWE. 1975. Freeze marking rodents with a pressurized refrigerant. Mammal Revue 5(1): 31.

LONG, CHARLES A. 1972. Taxonomic revision of the mammalian genus *Microsorex* Coues. Transactions of the Kansas Academy of Sciences 74: 181–196.

LORD, R. D. 1959. The lens as an indicator of age in the cottontail rabbit. Journal of Wildlife Management 23: 358–360.

LOWERY, GEORGE H., JR. 1974. The mammals of Louisiana and its adjacent waters. Louisiana State University Press, Baton Rouge. 565 pp.

MERRIAM, C. HART. 1884. The vertebrates of the Adirondack region: The Mammalia. Transactions Linnaean Society of New York 1: 1–214.

MILLER, G. S., JR., AND G. M. ALLEN. 1928. The American bats of the genera *Myotis* and *Pizonyx*. United States National Museum Bulletin 144. 218 pp.

MOHR, CHARLES E. 1933. Pennsylvania bats of the genus *Myotis*. Proceedings Pennsylvania Academy of Science 7: 39–43.

———. 1936. Notes on the least bat, *Myotis subulatus leibii*. Proceedings Pennsylvania Academy of Science 10: 62–65.

NELSON, E. W. 1930. Wild animals of North America. National Geographic Society, Washington, D.C. 254 pp.

NEWMAN, H. H. 1909. A case of normal identical quadruplets in the nine-banded armadillo, and its bearing on the problems of identical twins and of sex determination. Biological Bulletin 17(3): 181–187.

O'NEIL, TED. 1949. The muskrat in the Louisiana coastal marshes. Louisiana Department Wildlife and Fisheries. 152 pp.

OVERTON, W. SCOTT, AND DAVID E. DAVIS. 1969. Estimating the number of animals in wildlife populations. *In* Wildlife Management Techniques, The Wildlife Society, Washington, D.C., pp. 403–455.

POOLE, EARL L. 1932. A survey of the mammals of Berks County, Pennsylvania. Reading Public Museum and Art Gallery, Bulletin No. 13. 74 pp.

———. 1938. Notes on the breeding of *Lasiurus* and *Pipistrellus* in Pennsylvania. Journal of Mammalogy 19:249.

RICE, DALE W. 1957. Life history and ecology of *Myotis austroriparius* in Florida. Journal of Mammalogy 38: 15–32.

SAUNDERS, P. B. 1929. *Microsorex* in captivity. Journal of Mammalogy 10: 78–79.

SAUNDERS, W. E. 1932. Notes on the mammals of Ontario. Transactions Royal Canadian Institute 18: 271–309.

SCHMIDT, F. J. W. 1931. Mammals of western Clark County, Wisconsin. Journal of Mammalogy 12: 99–117.

SCHWARTZ, ALBERT. 1954. Oldfield mice, *Peromyscus polionotus*, of South Carolina. Journal of Mammalogy 35: 561–569.

SEVERINGHAUS, C. W. 1949. Tooth development and wear as criteria of age in the white-tailed deer. Journal of Wildlife Management 13: 195–216.

SHELDON, W. G. 1949. A trapping and tagging technique for wild foxes. Journal of Wildlife Management 13: 309–311.

SHERMAN, HARLEY B. 1930. Birth of the young of *Myotis austroriparius*. Journal of Mammalogy 38: 15–32.

———. 1935. Food habits of the seminole bat. Journal of Mammalogy 16: 223–224.

———. 1939. Notes on the food of some Florida bats. Journal of Mammalogy 20: 103–104.

SNYDER, L. L. 1924. Some details on the life history and behaviour of *Napaeozapus insignis abietorum*. Journal of Mammalogy 5: 233–237.

SOKOLOV, I. I. 1973. Trends of evolution and the classification of the subfamily Lutrinae (Mustelide, Fissipedia). Bulletin Moscow Prir. Biol. 78(6): 45–52.

SPITZER, NUMIC, AND JAMES D. LAZELL, JR. 1978. A new rice rat (Genus *Oryzomys*) from Florida's Lower Keys. Journal of Mammalogy 59: 789–792.

SPRINGER, STEWART. 1937. Observations on *Cryptotis floridana* in captivity. Journal of Mammalogy 18: 237–238.

STEGEMAN, LEROY. 1930. Notes on *Synaptomys cooperi cooperi* in Washtenaw County, Michigan. Journal of Mammalogy 11: 460–466.

STODDARD, HERBERT L. 1932. The bobwhite quail: Its habits, preservation and increase. Charles Scribner's Sons, New York.

SUMNER, FRANCIS B. 1926. An analysis of geographic variation in mice of the Peromyscus polionotus group from Florida and Alabama. Journal of Mammalogy 7: 149–184.

SVIHLA, ARTHUR. 1934. The mountain water shrew. Murrelet 15:44–45.

TAYLOR, D. F. 1971. A radio-telemetry study of the black bear (*Euarctos americanus*) with notes on its present history and present status in Louisiana. Unpublished Master's thesis, Louisiana State University, Baton Rouge. 87 pp.

VAN ZYLL DE JONG, C. G. 1972. A systematic review of the Nearctic and Neotropical river otters (genus *Lutra*, Mustelidae, Carnivora). Life Science Contribution of the Royal Ontario Museum, Toronto, 80: 1–104.

WADE, OTIS. 1930. The behaviour of certain spermophiles with special reference to aestivation and hibernation. Journal of Mammalogy 11: 160–188.

WHITAKER, JOHN O., JR. 1970. The biological subspecies: An adjunct of the biological species. The Biologist 52(1): 12–15.

———. 1974. *Cryptotis parva*. Mammalian species no. 43. American Society of Mammalogists. 8 pp.

WHITAKER, JOHN O., JR., AND L. L. SCHMELTZ. 1974. Food and external parasites of the eastern mole, *Scalopus aquaticus* from Indiana. Proceedings Indiana Academy of Sciences 83: 478–481.

WHITAKER, JOHN O., JR., AND D. D. PASCAL, JR. 1971. External parasites of arctic shrews taken in Pine County, Minnesota. Journal of Mammalogy, 52: 202.

WRIGHT, BRUCE. 1961. The latest specimen of the eastern puma. Journal of Mammalogy 42: 278–279.

GUIDE TO FURTHER READING

Pertinent references to various topics treated in this book are listed below. They include morphological, regional, systematic, and life history studies. The list is by no means exhaustive, but in the studies cited, additional references will be found that should prove useful to the reader who desires a fuller account of individual species.

General

ALLEN, GLOVER M. 1904. Check list of the mammals of New England. Fauna of New England. Occasional Papers Boston Society Natural History, 7(3): 1–35.

――――. 1942. Extinct and vanishing mammals of the Western Hemisphere, with the marine species of all the oceans. American Commission International Wildlife Protection Special Publication 11. N.Y. Zoological Park, New York. 620 pp.

AMERICAN MIDLAND NATURALIST. University of Notre Dame, Notre Dame, Ind.

ANDERSON, S., AND J. K. JONES, eds. 1967. Recent mammals of the world, a synopsis of families. Ronald Press Co., New York. 453 pp.

ANTHONY, H. E. 1928. Field book of North American mammals. Putnam, New York. 674 pp.

ARTHUR, STANLEY. 1928. The fur animals of Louisiana. Department of Conservation, New Orleans, Bulletin 18. 444 pp.

ASDELL, S. A. 1964. Patterns of mammalian reproduction. 2d ed. Comstock Publishing Associates, Ithaca, N.Y. 670 pp.

AUDUBON, J. J., AND J. BACHMAN. 1846–1854. The viviparous quadrupeds of North America. 3 vols. Privately published, New York.

BAILEY, J. W. 1946. Mammals of Virginia. Privately published, Richmond. 416 pp.

BAILEY, VERNON. 1923. The Mammals of the District of Columbia. Proceedings Biological Society Washington. 35 pp.

BANGS, OUTRAM. 1899. The land mammals of Peninsular Florida and the coast region of Georgia. Proceedings Boston Society Natural History 28: 157–235.

BARBOUR, R. W., AND W. H. DAVIS. 1974. Mammals of Kentucky. University of Kentucky Press, Lexington. 322 pp.

BOURLIERE, FRANCOIS. 1954. The natural history of mammals. Alfred A. Knopf, New York. 363 pp.

BRIMLEY, C. S. 1905. A descriptive catalogue of the mammals of North Carolina, exclusive of the Cetacea. J. Elisha Mitchell Science Society 21: 1–32.

BROOKS, D. M. 1959. Fur animals of Indiana. Department of Conservation Bulletin 4, Indianapolis. 195 pp.

BROOKS, FRED E. 1911. The mammals of West Virginia. Report West Virginia Board of Agriculture 20: 9–30.

BURT, WILLIAM H. 1946. Mammals of Michigan. University of Michigan Press, Ann Arbor. 288 pp.

――――. 1957. Mammals of the Great Lakes region. University of Michigan Press, Ann Arbor. 246 pp.

BURT, WILLIAM H., AND R. P. GROSSENHEIDER. 1976. A field guide to the mammals. 3d ed. Houghton Mifflin, Boston. 284 pp.

CONNOR, PAUL F. 1971. The mammals of Long Island, New York. New York State Museum and Science Service Bulletin 416. 78 pp.

——. The mammals of New York. New York State Museum and Science Service. In preparation.

CORY, CHARLES B. 1912. The Mammals of Illinois and Wisconsin. Field Museum Natural History. Zoological Series 11: 1–505.

CRONAN, J. M., AND A. BROOKS. 1968. The mammals of Rhode Island. Rhode Island Department of Natural Resources, Division Conservation of Wildlife Pamphlet No. 6. 133 pp.

DARLINGTON, PHILIP J., JR. 1957. Zoogeography: The geographical distribution of animals. John Wiley & Sons, New York. 675 pp.

DAVIS, J. W., AND R. C. ANDERSON, eds. 1971. Parasitic diseases of wild mammals. Iowa State University Press, Ames. 364 pp.

DEBLASE, A. F., AND R. E. MARTIN. 1974. A manual of mammalogy with keys to families of the world. Wm. C. Brown, Dubuque, Iowa. 329 pp.

DORAN, D. J. 1954–55. A catalogue of the Protozoa and helminths of North American rodents (in four parts). American Midland Naturalist 52: 118–128, 469–580; 53: 162–175, 446–454.

DOUTT, J. K., C. A. HEPPENSTALL, AND J. E. GUILDAY. 1967. The mammals of Pennsylvania. Pennsylvania Game Commission, Harrisburg. 281 pp.

DURCHER, B. H. 1903. Mammals of Mount Katahdin, Maine. Proceedings Biological Society of Washington 16: 63–71.

EISENTRAUT, M. 1956. Der Winterschlaf mit seinen ökologischen und physiologischen Begleiterscheinungen. VEB Gustav Fischer Verlag, Jena, Germany. 160 pp.

ERRINGTON, PAUL. 1946. Predation and vertebrate populations. Quarterly Review Biology 21(2): 144–177 and (3): 221–245.

GLASS, B. P. 1951. A key to the skulls of North American mammals. Burgess Publishing, Minneapolis. 53 pp.

GOLLEY, F. B. 1962. Mammals of Georgia. University of Georgia Press, Athens. 218 pp.

——. 1966. South Carolina mammals. Contributions Charleston Museum, 15. 181 pp.

GOODWIN, G. G. 1935. The mammals of Connecticut. State Geological and Natural History Survey Bulletin 53, Hartford. 221 pp.

GREENE, E. C. 1935. Anatomy of the rat. Transactions American Philosophical Society 27, Philadelphia. 370 pp.

GUNDERSON, HARVEY L. 1976. Mammalogy. McGraw-Hill, New York. 483 pp.

HAHN, WALTER LOUIS. 1909. The Mammals of Indiana. 33d Annual Report, Department Geology and Natural Resources Indiana, pp. 419–654.

HALL, E. R., AND K. R. KELSON. 1959. The mammals of North America. 2 vols. The Ronald Press, New York 1083 pp.

HAMILTON, W. J., JR. 1939. American mammals. McGraw-Hill, New York. 434 pp.

——. 1941. Notes on some mammals of Lee County, Florida. American Midland Naturalist 25: 686–691.

HANDLEY, C. O., JR. 1959. A revision of American bats of the genera *Euderma* and *Plecotus*. Proceedings of U.S. National Museum 110: 95–246.

HANDLEY, C. O., JR., AND C. P. PATTON. 1947. Wild mammals of Virginia. Virginia Commission Game and Inland Fish, Richmond. 220 pp.

HOFFMEISTER, D. F., AND C. O. MOHR. 1972. Fieldbook of Illinois Mammals. Illinois Natural History Survey Manual 4. Urbana. 233 pp.

HOWELL, A. B. 1926. Anatomy of the wood rat. Williams & Wilkins Co., Baltimore. 225 pp.

HOWELL, A. H. 1921. A biological survey of Alabama. I. Physiography and life zones. II. The mammals. North American Fauna No. 45. Washington, D.C., 88 pp.

JACKSON, C. F. 1922. Notes on New Hampshire mammals. Journal of Mammalogy 3: 13–15.

JONES, J. K., C. D. CARTER, AND H. H. GENOWAYS. 1975. Revised checklist of North American mammals north of Mexico. Occasional Papers Museum Texas Tech. University 28: 1–14.

JOURNAL OF MAMMALOGY. A quarterly published by the American Society of Mammalogists and a veritable storehouse of mammal lore. Everyone interested in mammals should have access to this journal.

KATZ, DAVID T. 1941. A key to the mammals of Ohio. Ohio Wildlife Research Station, Ohio State University. Mimeographed. 34 pp.

KAYSER, C. 1961. The physiology of natural hibernation. Pergamon Press, Oxford. 325 pp.

KENNEDY, MICHAEL L., KENNETH N. RANDOLPH, AND TROY L. BEST. 1974. A review of Mississippi mammals. Studies in Natural Science 2, No. 1. Natural Science Research Institute, Eastern New Mexico University. 36 pp.

KIRK, GEORGE L. 1916. The mammals of Vermont. Joint Bulletin No. 2, Vermont Botanical and Bird Clubs, pp. 28–34.

LAWLOR, T. E. 1976. Handbook to the orders and families of living mammals. Mad River Press, Eureka, California. 244 pp.

LAYNE, J. N. 1958. Notes on mammals of southern Illinois. American Midland Naturalist 60: 219–254.

LINZEY, A. V., AND D. W. LINZEY. 1971. Mammals of Great Smoky Mountains National Park. University of Tennessee Press. 114 pp.

LOWERY, GEORGE H., JR. 1974. The mammals of Louisiana and its adjacent waters. Louisiana State University Press, Baton Rouge. 565 pp.

LYON, MARCUS WARD, JR. 1936. Mammals of Indiana. American Midland Naturalist 17: 1–384.

LYMAN, C. P., AND A. R. DAWE. 1960. Mammalian hibernation. Bulletin of the Museum of Comparative Zoology 124. Harvard, Cambridge, Mass. 549 pp.

MAMMAL REVIEW. Mammal Society (England). Blackwell Scientific Publications. Oxford.

MAMMALIA. 55, rue de Buffon. Paris.

MARLER, PETER, AND WILLIAM J. HAMILTON III. 1966. Mechanisms of animal behavior. John Wiley and Sons, New York. 771 pp.

MARTIN, A. C., H. S. ZIM, AND A. L. NELSON. 1951. American wildlife and plants: A guide to wildlife food habits. Dover, New York 500 pp.

MATHEWS, L. HARRISON. 1971. The life of mammals. 2 vols. Universe Books, New York.

MERRIAM, C. HART. 1884. The vertebrates of the Adirondack region: The Mammalia. Transactions Linnaean Society of New York 1: 1–214.

MILLER, GERRIT S., JR. 1899. Preliminary list of the mammals of New York. Bulletin of N.Y. State Museum 6, no. 29.

MILLER, GERRIT S., JR., AND REMINGTON KELLOGG. 1955. List of North American recent mammals. U.S. National Museum Bulletin 205. 954 pp.

MILLER, M. E. 1952. Guide to the dissection of the dog. 3d ed. Edwards Bros., Ann Arbor, Mich. 369 pp.

MORRIS, DESMOND. 1965. The mammals: A guide to the living species. Hodder and Stoughton, London.

MUMFORD, R. E. 1969. Distribution of the mammals of Indiana. Indiana Academy of Sciences Monograph No. 1. 114 pp.

MUMFORD, R. E., AND J. O. WHITAKER, JR. (In press) The mammals of Indiana. Indiana University Press.

MURIE, O. J. 1954. A field guide to animal tracks. Houghton-Mifflin, Boston. 375 pp.

NECKER, WALTER L., AND DONALD M. HATFIELD. 1941. Mammals of Illinois. Bulletin Chicago Academy Sciences 6(3): 17–60.

NELSON, E. W. 1930. Wild animals of North America. National Geographic Society, Washington, D.C. 254 pp.

NORTON, ARTHUR H. 1930. Mammals of Portland, Maine and vicinity. Proceedings of the Portland Society Natural History 4(1): 1–151.

OSGOOD, FREDERICK L., JR. 1938. The mammals of Vermont. Journal of Mammology 19: 435–441.

PARADISO, J. L. 1969. Mammals of Maryland. North American Fauna No. 66. Washington, D.C. 193 pp.

PETERSON, RANDOLPH. 1966. The mammals of eastern Canada. Oxford University Press, Toronto. 465 pp.

RHOADS, SAMUEL N. 1896. Contributions to the biology of Tennessee, No. 3: Mammalia. Proceedings Academy Natural Sciences of Philadelphia 48: 175–205.

––––. 1903. The mammals of Pennsylvania and New Jersey. Privately published, Philadelphia. 266 pp.

ROWLANDS, I. W., ed. 1966. Comparative biology of reproduction in mammals. Academic Press, New York.

SEALANDER, JOHN A. 1979. A guide to Arkansas mammals. River Road Press, Conway, Arkansas. 313 pp.

SETON, E. T. 1909. Life histories of northern animals. Vol. 1. Grasseaters. Charles Scribner's Sons, New York. 673 pp.

––––. 1929. Lives of game animals. 4 vols. Doubleday, Doran & Co., New York.

SHERMAN, FRANKLIN. 1937. Some mammals of western South Carolina. Journal of Mammalogy 18: 512–513.

SHERMAN, H. B. 1936. A list of the recent land mammals of Florida. Proceedings of the Florida Academy of Science 1: 102–128.

SIMPSON, G. G. 1945. The principles of classification and a classification of mammals. Bulletin American Museum Natural History, New York, 85. 350 pp.

SOUTHWESTERN NATURALIST.

STEVENSON, H. M. 1976. Vertebrates of Florida: Identification and distribution. University Press of Florida, Gainesville. 607 pp.

VAN GELDER, RICHARD G. 1969. Biology of mammals. Charles Scribner's Sons, New York. 197 pp.

VAUGHN, T. A. 1972. Mammalogy. Saunders, Philadelphia. 463 pp.

WALKER, E. P., ed. 1968. Mammals of the world. 3 vols. The Johns Hopkins Press, Baltimore.

WARD, R. P. 1965. The mammals of Mississippi. Journal Mississippi Academy of Science 11: 309–330.

WHITAKER, JOHN O., JR. 1968. Keys to the vertebrates of the eastern United States, excluding birds. Burgess Publishing Co., Minneapolis. 256 pp.

WHITAKER, J. O., JR., AND N. WILSON. 1974. Host and distribution list of mites (Acari), parasitic and phoretic, in the hair of wild mammals of North America, north of Mexico. American Midland Naturalist 91: 1–67.

WOLFE, J. L. 1971. Mississippi land mammals. Mississippi Museum Natural Science, Jackson. 44 pp.

YOUNG, J. Z. 1957. The life of mammals. Oxford University Press, Oxford. 820 pp.

Marsupialia

FITCH, H. S., AND L. L. SANDRIDGE. 1953. Ecology of the opossum on a natural area in northeastern Kansas. University of Kansas, Publication Museum Natural History 7: 305–338.

GARDNER, A. L. 1973. The systematics of the genus *Didelphis* (Marsupialia: Didelphidae) in North and Middle America. Special Publication The Museum. Texas Tech. University, No. 4. 81 pp.

HARTMAN, C. G. 1952. Possums. University of Texas Press, Austin. 174 pp.

———. 1953. Breeding habits, development and birth of the opossum. Smithsonian Report for 1921. Pp. 347–363.

McCRADY, E., JR. 1938. The embryology of the opossum. The American Anatomical Memoirs No. 16. 233 pp.

Insectivora

ARLTON, A. V. 1936. An ecological study of the common mole. Journal of Mammalogy 17: 349–371.

BLOSSOM, P. M. 1932. A pair of long-tailed shrews (*Sorex cinereus*) in captivity. Journal of Mammalogy 13: 136–143.

CHOATE, J. R. 1970. Systematics and zoogeography of middle American shrews of the genus *Cryptotis*. University of Kansas, Publications of the Museum of Natural History. 19: 195–317.

CHRISTIAN, J. J. 1950. Behavior of the mole (*Scalopus*) and the shrew (*Blarina*). Journal of Mammalogy 31: 281–287.

CLOUGH, G. C. 1963. Biology of the arctic shrew, *Sorex arcticus*. American Midland Naturalist 69: 69–81.

CONAWAY, C. H. 1952. Life history of the water shrew, *Sorex palustris navigator*. American Midland Naturalist 48: 219–248.

———. 1958. Maintenance, reproduction and growth of the least shrew in captivity. Journal of Mammalogy 39: 507–512.

CROWCROFT, PETER. 1957. The life of the shrew. Max Reinhardt, London. 166 pp.

GENOWAYS, H. H., AND J. R. CHOATE. 1972. A multivariate analysis of systematic relationships among populations of the short-tailed shrew (genus *Blarina*) in Nebraska. Systematic Zoology 21: 106–116.

GOULD, E., N. C. NEGUS, AND A. NOVICK. 1964. Evidence for echolocation in shrews. Journal of Experimental Zoology 156: 19–38.

HAMILTON, W. J., JR. 1929. Breeding habits of the short-tailed shrew, *Blarina brevicauda*. Journal of Mammalogy 10: 125–134.

―――. 1930. The food of the Soricidae. Journal of Mammalogy 11: 26–39.

―――. 1934. Habits of *Cryptotis parva* in New York. Journal of Mammalogy 15: 154–155.

―――. 1944. The biology of the little short-tailed shrew, *Cryptotis parva*. Journal of Mammalogy 25: 1–7.

JACKSON, HARTLEY H. T. 1915. A review of the American moles. North American Fauna No. 38. Washington, D.C. 100 pp.

MERRIAM, C. HART. 1895. Revision of the shrews of the American genera *Blarina* and *Notiosorex*. Gerrit S. Miller, Jr. The long-tailed shrews of the eastern United States. C. Hart Merriam. Synopsis of the American shrews of the genus *Sorex*. North American Fauna No. 10. Washington, D.C. 124 pp.

PEARSON, O. P. 1942. On the cause and nature of a poisonous action produced by the bite of a shrew (*Blarina brevicauda*). Journal of Mammalogy 23: 159–166.

WHITAKER, J. O., JR., AND R. E. MUMFORD. 1972. Food and ectoparasites of Indiana shrews. Journal of Mammalogy 53: 329–335.

Chiroptera

ALLEN, GLOVER M. 1939. Bats. Harvard University Press, Cambridge, Mass. 368 pp.

BARBOUR, R. W., AND W. H. DAVIS. 1969. Bats of America. University of Kentucky Press, Lexington. 286 pp.

BEER, J. R. 1955. Survival and movements of banded big brown bats. Journal of Mammalogy 36: 242–248.

BEER, J. R., AND A. G. RICHARDS. 1956. Hibernation of the big brown bat. Journal of Mammalogy: 37: 31–41.

CHRISTIAN, J. J. 1956. The natural history of a summer aggregation of the big brown bat, *Eptesicus fuscus fuscus*. American Midland Naturalist 55: 66–95.

DAVIS, R. B., C. F. HERREID II, AND H. L. SHORT. 1962. Mexican free-tailed bats in Texas. Ecological Monographs 32: 311–364.

DAVIS, W. H., R. W. BARBOUR, AND M. D. HASSELL. 1968. Colonial behavior of *Eptesicus fuscus*. Journal of Mammalogy 49: 44–50.

DAVIS, W. H., AND W. Z. LIDICKER, JR. 1956. Winter range of the red bat, *Lasiurus borealis*. Journal of Mammalogy 37: 280–281.

EDGERTON, H. E., P. F. SPANGLE, AND J. K. BAKER. 1966. Mexican free-tail bats: Photography. Science 153: 201–203.

GATES, WILLIAM H. 1936. Keeping bats in captivity. Journal of Mammalogy 17: 268–273.

GRIFFIN, D. R. 1940. Migrations of New England bats. Bulletin of the Museum of Comparative Zoology 86(6): 217–246.

_____. 1958. Listening in the dark. Yale University Press, New Haven. 413 pp.

HALL, E. R., AND W. W. DALQUEST. 1950. A synopsis of the American bats of the genus *Pipistrellus*. University of Kansas, Publications Museum Natural History 1: 591–602.

HALL, E. R., AND J. K. JONES, JR. 1961. North American yellow bats, "*Dasypterus*," and a list of the named kinds of the genus *Lasiurus* Gray. University of Kansas, Publications Museum Natural History 14: 73–98.

HALL, J. S. 1962. A life history and taxonomic study of the Indiana bat, *Myotis sodalis*. Reading Public Museum and Art Gallery, Science Publication No. 12. 68 pp.

HUMPHREY, S. R., AND J. B. COPE. 1976. Population ecology of the little brown bat, *Myotis lucifugus*, in Indiana and north-central Kentucky. American Society of Mammalogists Special Publication No. 4. 81 pp.

LaVAL, R. K. 1970. Infraspecific relationships of bats of the species *Myotis austroriparius*. Journal of Mammalogy 51: 542–552.

McNAB, BRIAN K. 1974. The behavior of temperate cave bats in a subtropical environment. Ecology 55(50): 943–958.

MILLER, GERRIT S., JR. 1897. Revision of the North American bats of the family Vespertilionidae. North American Fauna No. 13. Washington, D.C. 1940. 135 pp.

_____. 1907. The families and genera of bats. U.S. National Museum Bulletin 57. 282 pp.

MILLER, GERRIT S., JR., AND G. M. ALLEN. 1928. The American bats of the genera *Myotis* and *Pizonyx*. U.S. National Museum Bulletin 144. 218 pp.

MOHR, C. E. 1933. Pennsylvania bats of the genus *Myotis*. Proceedings Pennsylvania Academy of Sciences 7: 39–43.

MURPHY, R. C., AND J. T. NICHOLS. 1913. Long Island Fauna and Flora: I. The Bats. Science Bulletin, Museum Brooklyn Institute of Arts and Sciences 2: 1–15.

PHILLIPS, G. L. 1966. Ecology of the big brown bat (Chiroptera: Vespertilionidae) in northeastern Kansas American Midland Naturalist 75: 168–198.

TUTTLE, M. D. 1974. Population ecology of the gray bat (*Myotis grisescens*): Factors influencing early growth and development. Occasional Papers, Museum National History, University of Kansas 36: 1–24.

WHITAKER, J. O., JR. 1972. Food habits of bats from Indiana. Canadian Journal of Zoology 50: 877–883.

_____. 1973. External parasites of bats in Indiana. Journal of Parasitology 59: 1148–1150.

WIMSATT, W. A., ed. 1970. Biology of Bats. Academic Press, New York & London. Vol. 1, 406 pp. Vol. 2, 477 pp. Vol. 3 (1977), 651 pp.

Edentata

KALMBACH, E. R. 1944. The armadillo: Its relation to agriculture and game. Game, Fish and Oyster Commission, Austin, Texas. 60 pp.

TALMAGE, R. V., AND G. D. BUCHANAN. 1954. The armadillo (*Dasypus novemcinctus*), a review of its natural history, ecology, anatomy and reproductive physiology. The Rice Institute Pamphlet 41. 135 pp.

Lagomorpha

ALDOUS, C. M. 1937. Notes on the life history of the snowshoe hare. Journal of Mammalogy 18: 46–57.

FAY, F. H., AND E. H. CHANDLER. 1955. The geographical and ecological distribution of cottontail rabbits in Massachusetts. Journal of Mammalogy 36: 415–424.

GRANGE, WALLACE B. 1932. The pelages and color changes of the snowshoe hare, *Lepus americanus phaeonotus* Allen. Journal of Mammalogy 13: 99–116.

HUNT, T. P. 1959. Breeding habits of the swamp rabbit with notes on its life history. Journal of Mammalogy 40: 82–91.

LECHLEITNER, R. R. 1959. Sex ratio, age classes and reproduction of the black-tailed jackrabbit. Journal of Mammalogy 40: 63–81.

LOWE, C. E. 1958. Ecology of the swamp rabbit in Georgia. Journal of Mammalogy 39: 116–127.

MACLULICH, D. A. 1937. Fluctuations in the numbers of the varying hare (*Lepus americanus*). Biology Series No. 43, University of Toronto Studies. Pp. 5–136.

NELSON, E. W. 1909. The rabbits of North America. North American Fauna No. 29. Washington, D.C. 314 pp.

SVIHLA, RUTH DOWELL. 1929. Habits of *Sylvilagus aquaticus littoralis*. Journal of Mammalogy 10: 315–319.

TERREL, T. L. 1972. The swamp rabbit (*Sylvilagus aquaticus*) in Indiana. American Midland Naturalist 87: 283–295.

TOMKINS, IVAN R. 1935. The marsh rabbit: An incomplete life history. Journal of Mammalogy 16: 201–205.

Rodentia

ALLEN, DURWARD L. 1942. Populations and habits of the fox squirrel in Allegan County, Michigan. American Midland Naturalist 27: 338–379.

_____. 1943. Michigan fox squirrel management. Department of Conservation, Michigan.

ALLEN, ELSA G. 1938. The habits and life history of the eastern chipmunk *Tamias striatus lysteri*. N.Y. State Museum Bulletin 314: 7–119.

BAILEY, VERNON. 1900. Revision of the American voles of the genus *Microtus*. North American Fauna No. 17. Washington, D.C. 88 pp.

_____. 1927. Beaver habits and experiments in beaver control. U.S. Department of Agriculture, Technical Bulletin. Pp. 1–39.

BAILEY, V., AND J. K. DOUTT. 1942. Two new beavers from Labrador and New Brunswick. Journal of Mammalogy 23: 86–88.

BENTON, A. H. 1955. Observations on the life history of the northern pine mouse. Journal of Mammalogy 36: 52–62.

BURT, WILLIAM H. 1940. Territorial behavior and populations of some small mammals in southern Michigan. University of Michigan, Museum of Zoology Miscellaneous Publication No. 45: 7–58.

CALHOUN, J. B. 1966. The ecology and sociology of the Norway rat. United States Department of Health, Education and Welfare.

CONNOR, PAUL F. 1959. The bog lemming *Synaptomys cooperi* in southern New Jersey. Michigan State University Publication Museum Series I: 161–248.

DAVIS, DAVID E. 1953. The characteristics of rat populations. Quarterly Review of Biology 28: 373–401.

ELTON, CHARLES. 1942. Voles, mice and lemmings. Oxford University Press, New York. 496 pp.

EVANS, F. C. 1951. Notes on a population of the striped ground squirrel (*Citellus tridecemlineatus*) in an abandoned field in southeastern Michigan. Journal of Mammalogy 32: 437–449.

EVANS, J. 1970. About Nutria and their control. Source Publication No. 86. Bureau Sport Fisheries and Wildlife, Denver. 65 pp.

FORD, S. D. 1977. Range, distribution and habitat of the western harvest mouse, *Reithrodontomys megalotis,* in Indiana. American Midland Naturalist 98: 422–432.

GOLDMAN, E. A. 1910. Revision of the wood rats of the genus *Neotoma.* North American Fauna No. 31. Washington, D.C. 124 pp.

_____. 1918. The rice rats of North America (genus *Oryzomys*). North American Fauna No. 43. Washington, D.C. 100 pp.

GOODPASTER, W. W., AND D. F. HOFFMEISTER. 1954. Life history of the golden mouse. *Peromyscus nuttalli,* in Kentucky. Journal of Mammalogy 35: 16–27.

GRIZZELL, R. A., JR. 1955. A study of the southern woodchuck, *Marmota monax monax.* American Midland Naturalist 53: 257–293.

HAMILTON, W. J., JR. 1935. Habits of jumping mice. American Midland Naturalist 16: 187–200.

_____. 1938. Life history notes on the northern pine mouse. Journal of Mammalogy 19: 163–170.

_____. 1939. Observations on the life history of the red squirrel in New York. American Midland Naturalist 22: 732–745.

_____. 1940. Life and habits of field mice. Scientific Monthly 50: 425–434.

_____. 1941. The food of small forest mammals in eastern United States. Journal of Mammalogy 22: 250–263.

_____. 1946. Habits of the swamp rice rat, *Oryzomys palustris palustris* (Harlan). American Midland Naturalist 36: 730–736.

_____. 1953. Reproduction and young of the Florida wood rat, *Neotoma f. floridana* (Ord.). Journal of Mammalogy 34: 180–189.

HATT, R. T. 1929. The red squirrel: Its life history and habits. Roosevelt Wild Life Annals, N.Y. State College Forestry, 2(1): 3–146.

HOLLISTER, N. 1911. A systematic synopsis of the muskrats. North American Fauna No. 32. Washington, D.C. 47 pp.

HOWARD, W. E. 1949. Dispersal, amount of inbreeding, and longevity in a local population of prairie deer mice on the George Reserve, southern Michigan. University of Michigan Contributions Vertebrate Biology 43. 50 pp.

HOWELL, A. B. 1927. Revision of the American lemming mice (genus *Synaptomys*). North American Fauna No. 50. Washington, D.C. 38 pp.

HOWELL, A. H. 1914. Revision of the American harvest mice (genus *Reithrodontomys*). North American Fauna No. 36. Washington, D.C. 97 pp.

_____. 1915. Revision of the American marmots. North American Fauna No. 37. Washington, D.C. 80 pp.

_____. 1918. Revision of the American flying squirrels. North American Fauna No. 44. Washington, D.C. 64 pp.

_____. 1938. Revision of the North American ground squirrels. North American Fauna No. 56. Washington, D.C. 256 pp.

JAMESON, E. W., JR. 1947. Natural history of the prairie vole. University of Kansas, Publications Museum Natural History 1: 125-151.

JOHNSON, CHARLES E. 1925. The muskrat in New York: Its natural history and economics. Roosevelt Wild Life Bulletin, 3: 199-320.

———. 1927. The beaver in the Adirondacks. Roosevelt Wild Life Bulletin 4: 501-641.

JOHNSON, M. S. 1926. Activity and distribution of certain wild mice in relation to biotic communities. Journal of Mammalogy 7: 245-277.

KING, J. A., ed. 1968. Biology of *Peromyscus* (Rodentia). Special Publication No. 2. American Society of Mammalogists 593 pp.

LAYNE, J. N. 1963. A study of the parasites of the Florida mouse, *Peromyscus floridanus*, in relation to host and environmental factors. Tulane Studies Zoology 11: 1-27.

LERAAS, HAROLD J. 1938. Observations on the growth and behavior of harvest mice. Journal of Mammalogy 19: 441-444.

LINSDALE, JEAN. 1927. Notes on the life history of *Synaptomys*. Journal of Mammalogy 8: 51-54.

LINZEY, D. W. 1968. An ecological study of the golden mouse, *Ochrotomys nuttalli*, in the Great Smoky Mountains National Park. American Midland Naturalist 79: 320-345.

LOWERY, G. H., JR., AND W. B. DAVIS. 1942. A revision of the fox squirrels of the Lower Mississippi Valley and Texas. Occasional Papers of the Museum of Zoology, Louisiana State University, No. 9: 153-172.

McCARLEY, W. H. 1958. Ecology, behavior and population dynamics of *Peromyscus nuttalli* in eastern Texas. Texas Journal Science 10: 147-171.

MARTIN, E. P. 1956. A population study of the prairie vole (*Microtus ochrogaster*) in northeastern Kansas. University Kansas, Publications, Museum Natural History 8: 361-416.

MERRIAM, C. HART. 1895. Monographic revision of the pocket gophers, Family Geomyidae (exclusive of the species of *Thomomys*). North American Fauna No. 8. Washington, D.C. 258 pp.

MILLER, GERRIT S., JR. 1896. The genera and subgenera of voles and lemmings. North American Fauna No. 12. Washington, D.C. 84 pp.

MOORE, J. C. 1957. The natural history of the fox squirrel, *Sciurus niger shermani*. Bulletin American Museum of Natural History 113: 1-71.

MORGAN, L. H. 1868. The American beaver and his works. Lippincott, Philadelphia. 330 pp.

NEGUS, N. C., E. GOULD, AND R. K. CHIPMAN. 1961. Ecology of the rice rat, *Oryzomys palustris* (Harlan), on Breton Island, Gulf of Mexico, with a critique of social stress theory. Tulane Studies Zoology 8: 93-123.

ODUM, E. P. 1955. An eleven-year history of a *Sigmodon* population. Journal of Mammalogy 36: 368-378.

OSGOOD, W. H. 1909. Revision of the mice of the American genus *Peromyscus*. North American Fauna No. 28. Washington, D.C. 285 pp.

PACKARD, R. L. 1956. The tree squirrels of Kansas. Ecology and economic importance. University of Kansas Museum Natural History and State Biological Survey, Kansas Miscellaneous Publication 11. 67 pp.

———. 1969. Taxonomic review of the golden mouse, *Ochrotomys nuttalli*. University of Kansas Museum Natural History Miscellaneous Publication 51: 373-406.

PEARSON, P. G. 1952. Observations concerning the life history and ecology of the woodrat, *Neotoma floridana floridana* (Ord.). Journal of Mammalogy 33: 459–463.

POOLE, EARL L. 1940. A life history sketch of the Allegheny woodrat. Journal of Mammalogy 21: 249–270.

PREBLE, E. A. 1899. Revision of the jumping mice of the genus *Zapus*. North American Fauna No. 15. Washington, D.C. 42 pp.

QUIMBY, D. C. 1951. The life history and ecology of the jumping mouse. *Zapus hudsonius*. Ecological Monographs 21: 61–95.

RAINEY, D. G. 1956. Eastern wood rat, *Neotoma floridana*: Life history and ecology. University of Kansas, Museum Natural History Publication 8: 535–645.

SCHOOLEY, J. P. 1934. A summer breeding season in the eastern chipmunk, *Tamias striatus*. Journal of Mammalogy 15: 194–196.

SCHWARTZ, A., AND E. P. ODUM. 1957. The woodrats of the eastern United States. Journal of Mammalogy 38: 197–206.

SHELDON, CAROLYN. 1938. Vermont jumping mice of the genus *Napaeozapus*. Journal of Mammalogy 19: 444–453.

SOLLBERGER, DWIGHT E. 1940. Notes on the life history of the small eastern flying squirrel. Journal of Mammalogy 21: 282–293.

STRUTHERS, PARKE H. 1928. Breeding habits of the Canadian porcupine (*Erethizon dorsatum*). Journal of Mammalogy 9: 300–308.

SUMNER, F. B., AND J. J. KAROL. 1929. Notes on the burrowing habits of *Peromyscus polionotus*. Journal of Mammalogy 10: 213–215.

SVIHLA, ARTHUR. 1931. Life history of the Texas rice rat (*Oryzomys palustris texensis*). Journal of Mammalogy 12: 238–242.

——. 1932. A comparative life history study of the mice of the genus *Peromyscus*. University Michigan, Museum Zoology Miscellaneous Publications 24: 6–39.

SVIHLA, ARTHUR AND RUTH DOWELL SVIHLA. 1931. The Louisiana muskrat. Journal of Mammalogy 12: 12–28.

SVIHLA, RUTH DOWELL. 1930. Notes on the golden harvest mouse. Journal of Mammalogy 11: 53–54.

TEVIS, L., JR. 1950. Summer behavior of a family of beavers in New York state. Journal of Mammalogy 31: 40–65.

UHLIG, H. G. 1956. The gray squirrel in West Virginia. Conservation Commission of West Virginia. 83 pp.

WADE, OTIS. 1927. Breeding habits and early life of the thirteen-striped ground squirrel, *Citellus tridecemlineatus* (Mitchill). Journal of Mammalogy 8: 269–276.

WHITAKER, J. O., JR. 1963. A study of the meadow jumping mouse, *Zapus hudsonius* (Zimmermann) in central New York. Ecological Monographs 33: 215–254.

——. 1966. Food of *Mus musculus, Peromyscus maniculatus bairdii* and *Peromyscus leucopus* in Vigo County, Indiana. Journal of Mammalogy 47: 473–486.

——. 1972. *Zapus hudsonius*. Mammalian species No. 11. American Society of Mammalogists. Pp. 1–7.

——. 1972. Food and external parasites of *Spermophilus tridecemlineatus* in Vigo County, Indiana. Journal of Mammalogy 53: 644–648.

WHITAKER, J. O., JR., AND R. E. MUMFORD. 1972. Ecological studies on *Reithrodontomys megalotis* in Indiana. Journal of Mammalogy 53: 850–860.

WING, E. S. 1960. Reproduction in the pocket gopher in North Central Florida. Journal of Mammalogy 41: 35–43.

WORTH, C. B. 1950. Field and laboratory observations on roof rats, *Rattus rattus* (Linnaeus), in Florida. Journal of Mammalogy 31: 293–304.

WRIGLEY, R. E. 1972. Systematics and biology of the woodland jumping mouse, *Napaeozapus insignis*. Illinois Biological Monographs 47. University of Chicago Press, Urbana. 117 pp.

YERGER, R. W. 1953. Home range, territoriality, and populations of the chipmunk in central New York. Journal of Mammalogy 34: 448–458.

ZIMMERMAN, E. G. 1965. A comparison of the habitat and food of two species of *Microtus*. Journal of Mammalogy 46: 605–612.

Carnivora

ALLEN, DURWARD L. 1939. Winter habits of Michigan skunks. Journal of Wildlife Management 3: 212–228.

BEKOFF, M., ed. 1978. Coyotes. Biology, behavior and management. Academic Press, New York. 384 pp.

COOK, D. B., AND W. J. HAMILTON, JR. 1957. The forest, the fisher and the porcupine. Journal of Forestry 55: 719–722.

COTTAM, C., A. L. NELSON, AND T. E. CLARKE. 1939. Notes on early winter food habits of the black bear in George Washington National Forest. Journal of Mammalogy 20: 310–314.

COUES, E. 1877. Fur-bearing animals: A monograph of the North American Mustelidae. Department of the Interior, U.S. Geological Survey of the Territories, Miscellaneous Publication No. 8. 348 pp.

CRABB, W. D. 1948. The ecology and management of the prairie spotted skunk in Iowa. Ecological Monographs 18: 201–232.

DEARBORN, N. 1932. Foods of some predatory fur-bearing animals of Michigan. University of Michigan, School of Forestry and Conservation Bulletin No. 1: 1–52.

EADIE, W. R., AND W. J. HAMILTON, JR. 1958. Reproduction in the fisher in New York. N.Y. Fish and Game Journal 5: 77–83.

ENDERS, R. K. 1952. Reproduction in the mink (*Mustela vison*). Proceedings American Philosophical Society 96: 691–755.

ERRINGTON, PAUL L. 1937. Summer food habits of the badger in northwestern Iowa. Journal of Mammalogy 1937: 213–216.

———. 1943. An analysis of mink predation upon muskrats in north-central United States. Iowa Agricultural Experiment Station Research Bulletin 320: 798–924.

GERSTELL, RICHARD. 1939. The growth and size of Pennsylvania black bears. Pennsylvania Game News. November, pp. 4–7.

GOLDMAN, E. A., AND H. H. T. JACKSON. 1950. Raccoons of North and Middle America. USDI. North American Fauna No. 60. Washington, D.C. 153 pp.

HALL, E. R. 1951. American weasels. University of Kansas, Publications Museum Natural History 4: 1–466.

HAMILTON, W. J., JR. 1933. The weasels of New York: Their natural history and economic status. American Midland Naturalist 14: 289–373.

———. 1935. Notes on food of red foxes in New York and New England. Journal of Mammalogy 16: 16–21.

_____. 1936. The food and breeding habits of the raccoon. Ohio Journal Science 36: 131–140.

_____. 1936. Seasonal food of skunks in New York. Journal of Mammalogy 17: 240–246.

_____. 1937. Winter activity of the skunk. Ecology 18: 326–327.

_____. 1940. The summer food of minks and raccoons on the Montezuma Marsh, New York. Journal of Wildlife Management 4: 80–84.

HAMILTON, W. J., JR., AND W. R. EADIE. 1964. Reproduction in the otter, *Lutra canadensis*. Journal of Mammalogy 45: 242–252.

HAMILTON, W. J., JR., AND RUSSELL P. HUNTER. 1939. Fall and winter food habits of Vermont bobcats. Journal Wildlife Management 3: 99–103.

HANSSON, A. 1947. The physiology of reproduction in mink (*Mustela vison*) with special reference to delayed implantation. Acta Zoologica 28: 1–137.

HOLLISTER, N. 1913. A synopsis of the American minks. Proceedings U.S. National Museum 44: 471–480.

HOWELL, A. H. 1901. Revision of the skunks of the genus *Chincha (Mephitis)*. North American Fauna No. 20. Washington, D.C. 62 pp.

_____. 1906. Revision of the skunks of the genus *Spilogale*. North American Fauna No. 26. Washington, D.C. 55 pp.

JOHNSON, A. S. 1970. Biology of the raccoon (*Procyon lotor varius* Nelson and Goldman) in Alabama. Auburn University Agriculture Experiment Station. 148 pp.

KORSCHGEN, L. J. 1948. December food habits of mink in Missouri. Journal of Mammalogy 39: 521–527.

LATHAM, R. M. 1950. The food of predaceous animals of northeastern United States. Pennsylvania Game Commission, Harrisburg. 69 pp.

LAYNE, J. N., AND N. M. MCCAULEY. 1977. Biological overview of the Florida Panther. In P. C. H. Pritchard, ed., Proceedings of the Florida Panther Conference. Florida Game and Freshwater Fish Commission, Gainesville. Pp. 5–45.

LIERS, E. E. 1951. Notes on the river otter (*Lutra canadensis*). Journal of Mammalogy 32: 1–9.

MATSON, J. R. 1954. Observations on the dormant phase of a female black bear. Journal of Mammalogy 35: 28–35.

MECH, L. DAVID. 1966. The wolves of Isle Royale. Fauna Parks U.S., Fauna Series 7. 210 pp.

_____. 1970. The wolf: The ecology and behaviour of an endangered species. The Natural History Press, Garden City, N.Y. 385 pp.

MENGEL, R. M. 1971. A study of dog-coyote hybrids and implications concerning hybridization in *Canis*. Journal of Mammalogy 52: 316–336.

MORSE, MARIUS A. 1937. Hibernation and breeding of the black bear. Journal of Mammalogy 18: 460–465.

MURIE, A. 1944. The wolves of Mt. McKinley. U.S.D.I. National Park Service, Fauna Series 5. 238 pp.

NELSON, A. L. 1933. A preliminary report on the winter food of Virginia foxes. Journal of Mammalogy 14: 40–43.

POLLACK, E. M. 1950. Breeding habits of the bobcat in northeastern United States. Journal of Mammalogy 31: 327–330.

——. 1951. Food habits of the bobcat in New England states. Journal of Wildlife Management 15: 209-209-213.

ROLLINGS, C. T. 1945. Habits, foods and parasites of the bobcat in Minnesota. Journal of Wildlife Management 9: 131-145.

SCOTT, T. G., AND W. D. KLIMSTRA. 1955. Red foxes and a declining prey population. Southern Illinois University, Monograph Series No. 1. 123 pp.

SEAGEARS, C. 1944. The fox in New York. N.Y. State Conservation Department, Albany. 85 pp.

SHAW, W. T. 1928. The spring and summer activities of the dusky skunk in captivity. N.Y. State Museum Handbook, Albany, 4: 5-103.

SHELDON, W. G. 1949. Reproductive behavior of foxes in New York State. Journal of Mammalogy 30: 236-246.

——. 1950. Denning habits and home range of red foxes in New York State. Journal of Wildlife Management 14: 33-42.

STUEWER, F. W. 1942. Raccoons, their habits and management in Michigan. Ecological Monographs 13: 203-257.

SULLIVAN, E. G. 1956. Gray fox reproduction, denning, range, and weights in Alabama. Journal of Mammalogy 37: 346-351.

SVIHLA, ARTHUR. 1931. Habits of the Louisiana mink (*Mustela vison vulgivagus*). Journal of Mammalogy 12: 366-368.

VAN GELDER, R. G. 1959. A taxonomic revision of the spotted skunks (genus *Spilogale*). Bulletin American Museum Natural History 117: 229-392.

VERTS, B. J. 1967. The biology of the striped skunk. University of Illinois Press, Urbana and Chicago. 218 pp.

WHITNEY, L. F., AND A. B. UNDERWOOD. 1952. The raccoon. Practical Science Publishing Company, Orange, Connecticut.

WRIGHT, P. L. 1942. Delayed implantation in the long-tailed weasel (*Mustela frenata*), the short-tailed weasel (*Mustela cicognani*), and the marten (*Martes americana*). Anatomical Record 83: 341-353.

WRIGHT, P. L., AND R. RAUSCH. 1955. Reproduction in the wolverine, *Gulo gulo*. Journal of Mammalogy 36: 346-355.

YEAGER, L. E. 1938. Otters of the delta hardwood region of Mississippi. Journal of Mammalogy 19: 195-201.

YOUNG, S. P. 1958. The bobcat of North America. Stackpole Co., Harrisburg, Pa.

YOUNG, S. P., AND E. A. GOLDMAN. 1944. The wolves of North America. American Wildlife Institute, Washington, D.C. 636 pp.

YOUNG, S. P., AND E. A. GOLDMAN. 1946. The puma, mysterious American cat. Stackpole Co., Harrisburg, Pa. 358 pp.

YOUNG, S. P., AND H. H. T. JACKSON. 1951. The clever coyote. Stackpole Co., Harrisburg, Pa. 411 pp.

Artiodactyla

BARBOUR, T., AND G. M. ALLEN. 1922. The white-tailed deer of eastern United States. Journal of Mammalogy 3: 65-78.

GOLDMAN, E. A., AND R. KELLOGG. 1940. Ten new white-tailed deer from North and Middle America. Proceedings Biological Society of Washington 53: 81–90.

MURIE, A. 1934. The moose of Isle Royale. University of Michigan, Museum Zoology Miscellaneous Publication 25: 7–44.

PETERSON, RANDOLPH. 1955. North American moose. University of Toronto Press. 280 pp.

TAYLOR, W. P. 1956. The deer of North America. The white-tailed, mule and black-tailed deer, genus *Odocoileus,* their history and management. Stackpole Company and Wildlife Management Institute. 668 pp.

TOWNSEND, M. T., AND M. W. SMITH. 1933. Part I. Townsend and Smith. The White-tailed deer of the Adirondacks. Part II. Charles J. Spiker. Some late winter and early spring observations on the white-tailed deer of the Adirondacks. Roosevelt Wildlife Bulletin 6(2): 161–325, 327–385.

Index

Mammals of the
Eastern United States

Designed by G. T. Whipple, Jr.
Composed by The Composing Room of Michigan, Inc.
in 10 point VIP Times Roman, 2 points leaded,
with display lines in Times Roman.
Printed offset by Vail-Ballou Press
on 70 pound Warren's Patina Coated Matte.
Bound by Vail-Ballou Press
in Holliston book cloth
and stamped in All Purpose foil.

Library of Congress Cataloging in Publication Data

HAMILTON, WILLIAM JOHN, 1902–
    Mammals of the Eastern United States.

    (Handbooks of American natural history)
    Bibliography: p.
    Includes index.
        1. Whitaker, John O., joint author.   II. Title.
III.   Series.
QL717.H3   1979   599'.097   79-12920
ISBN   0-8014-1254-4